The Vandana Shiva Reader

THE

VANDANA SHIVA

READER

VANDANA SHIVA

Foreword by
WENDELL BERRY

UNIVERSITY PRESS OF KENTUCKY

Published by The University Press of Kentucky

Scholarly publisher for the Commonwealth,
serving Bellarmine University, Berea College, Centre College of Kentucky, Eastern
Kentucky University, The Filson Historical Society, Georgetown College, Kentucky
Historical Society, Kentucky State University, Morehead State University, Murray
State University, Northern Kentucky University, Transylvania University, University
of Kentucky, University of Louisville, and Western Kentucky University.
All rights reserved.

Editorial and Sales Offices: The University Press of Kentucky
663 South Limestone Street, Lexington, Kentucky 40508-4008
www.kentuckypress.com

Figures by Dick Gilbreath

Library of Congress Cataloging-in-Publication Data

Shiva, Vandana, author.
  The Vandana Shiva reader / Vandana Shiva ; foreword by Wendell Berry.
    pages cm. — (Culture of the land)
  Includes bibliographical references and index.
  ISBN 978-0-8131-4560-0 (hardcover : alk. paper) —
  ISBN 978-0-8131-5329-2 (pbk. : alk. paper) —
  ISBN 978-0-8131-4699-7 (pdf : alk. paper) —
  ISBN 978-0-8131-4698-0 (epub : alk. paper)
  1. Shiva, Vandana. 2. Agricultural literature. I. Title. II. Series: Culture of the land.
  S494.5.A39S45 2015
  630—dc23   2014043937

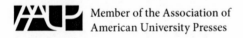

Contents

Foreword

Wendell Berry

I count it a privilege and a pleasure to be Vandana Shiva's friend—not least because, as her friend, I am spared the pain and suffering that she bestows upon her enemies. Her enemies are the radical oversimplifiers, the colonizers, of the global industrial corporations and their for-hire experts; the complacent, the indifferent, the inert; and the do-gooders who think that any version or device of technological progress is a charity.

Her advocacy for many years has been "the defense of the local through a global alliance." Nobody better represents that possibility or has done more to promote it. In support of her friends and allies in that necessary effort, she has probably spent more time in the air than most aviators. She has been an indispensable ally: devoted, smart, informed, tireless, and fearless.

Her great virtue as an advocate is that she is not a reductionist. Her awareness of the complex connections among economy and nature and culture preserves her from oversimplification. So does her understanding of the importance of diversity.

Like many people, she is against monoculture in farming. Like too few, she knows how monocultures on the land proceed from monoculture in the mind. Mental monoculture is the typical product of the modern university. Ananda Coomaraswamy was thinking of this when he said that people who spend four years in getting a good university education must then spend forty years in getting over it.

I said that Vandana is smart, and I think that is the right word for her. But I need to be more exact. Beyond what now passes for education, she has the experience, the good sense, and the formal intelligence necessary to understand both the complex local structures of traditional farming and forestry and the necrotic ramifications of global industrialism.

The impulse of her work comes, as we might expect, from her profound sympathy with the diverse small farmers of the third world, the majority of whom are women. But let us remember that the farm and forest landscapes of the "developed" nations are suffering the same colonialism as the third world. The subjection is everywhere the same, and with the same global purpose of making producers and consumers completely the hostages of corporations supplying seed, chemicals, machines, and fuel. Vandana has understood this industrial totalitarianism exactly, and she speaks for us all.

From Quanta to the Seed

An Unpredictable Journey

Physics was my passion and my chosen profession. In school I received the Science Talent Scholarship, which gave me the opportunity to train in India's leading scientific institutions. I trained to be a nuclear physicist in the Baba Atomic Research Centre, but moved to theoretical physics when my sister Mira, a medical doctor, made me aware of nuclear hazards. I realized then that most science is partial. I wanted to practice a holistic science and was drawn to quantum theory for its nonreductionist, nonmechanist paradigm.

Before leaving for Canada to do my PhD in the foundations of quantum theory, I wanted to visit my favorite places in the Himalaya where I had grown up. But the forests and streams had disappeared: the rich oak forests, which absorb the monsoon rains to release the water slowly as streams, had been cut down in the insane rush to build dams and roads, and to grow apples.

In looking for a way to deal with this personally experienced loss, I become a volunteer for the Chipko Movement—the movement to embrace trees to prevent their being cut. Every vacation from 1974 to 1981 found me back in the Himalayan village of Tehri Garhwal and often in the ashram of Bimla and Sunderlal Bahuguna, Gandhians who played a leading role in supporting the women who had spontaneously started the movement. By 1981 the government was forced to recognize the wisdom of Chipko—that the primary products of the forest are soil, water, and pure air, not timber, resin, and revenues. Logging was banned in the Himalaya above one thousand meters.

Though I had grown up in the forests of the Himalaya, it was the Chipko Movement that awakened my ecological consciousness and made me deeply aware of the relationship between ecological destruction and the creation of poverty. I often say that I have learned my ecology in "Chipko University," with ordinary peasant women as my teachers. "Staying Alive" was born of those teachings. One of the slogans that emerged from Chipko was that forestry should shift from commercial forestry to social forestry—in the service of nature and society.

I returned to India after receiving my PhD because I wanted to give back to my society and also understand it better. It would have been easy to get a tenure-track position in a North American university, but I chose the more difficult and challenging path of trying to combine scientific research with social and ecological responsibility.

In 1980, while I was at the Indian Institute of Management (IIM) in Bangalore, we witnessed the conversion of farmlands around the city into monocultures of eucalyptus plantations. In trying to understand what was driving this, we found the World Bank was financing eucalyptus plantations for the pulp industry, calling it "social forestry." *Ecological Audit of Eucalyptus,* a study I and my colleagues at IIM conducted, had a huge impact. It traveled to Thailand, Portugal, and Brazil and was used to challenge the idea that commercial plantations are forests. I realized that humanity had cultivated a "monoculture of the mind," which created a blindness to diversity and its potentials, a blindness that blocked out the high productivity of biodiverse systems in forests, in agriculture, in the ocean.

It was becoming increasingly evident that scientific expertise worked more in the service of capital and forces abetting the destruction of nature. I wanted to work in the service of people and nature. In 1981, I took the decision to leave academics and start an independent institute to support grassroots ecology movements. My mother gave me her cowshed, and the Research Foundation for Science, Technology, and Ecology was born. We shut down mines in Doon Valley; we assessed the impacts of dams and power plants in Narmada, Suverna Rekha, Tehri, and Singrauli.

Mr. S. M. Mohamed Idris, the founder of the Consumers Association of Penang and Third World Network, Malaysia, saw the study and invited me to a forest conference in Penang. Out of that meeting came the World Rainforest Movement, the network of forest defenders worldwide.

In 1984, a number of tragic events took place in India. In June, the Golden Temple was attacked because it was harboring extremists. In October, Indira Gandhi was assassinated. And in December, a terrible industrial disaster took place in Bhopal when Union Carbine's pesticide plant leaked a toxic gas. Thirty thousand people died in the terrorism in Punjab, and thirty thousand people have died in the "industrial terrorism" of Bhopal. This is equivalent to twelve 9/11s. I was forced to sit up and ask why agriculture had become like war. Why did the "Green Revolution," which had received the Nobel Peace Prize, breed extremism and terrorism in Punjab? This questioning led to my books *The Violence of the Green Revolution* and *Monocultures of the Mind*. Blindness to diversity and self-organization in nature and society was clearly a basic problem in the mechanistic, Cartesian industrial paradigm. And this blindness led to false claims that industrial monocultures in forestry, farming, fisheries, and animal husbandry produced more food and were necessary to alleviate hunger and poverty. On the contrary, monocultures produce less and use more inputs, thus destroying the environment and impoverishing people.

In 1987, the Dag Hammarjold Foundation organized a meeting on biotechnology in Geneva called Laws of Life. I was invited because of my book on the Green Revolution. At the conference, the biotech industry laid out its plans—to patent life; to genetically engineer seeds, crops, and life-forms; and to get full freedom to trade through the GATT negotiations, which finally led to the WTO. This led to my focus on intellectual property rights, free trade, globalization—and to a life dedicated to saving seeds and promoting organic farming as an alternative to a world dictated and controlled by corporations.

Having dedicated my life to the defense of the intrinsic worth of all species, the idea of life-forms, seeds, and biodiversity being reduced to corporate inventions and hence corporate property was abhorrent to me. Further, if seeds become "intellectual property," saving and sharing seeds become intellectual property theft. Our highest duty, to save seeds, becomes a criminal act. The legalizing of the criminal act of owning and monopolizing life through patents on seeds and plants was morally and ethically unacceptable to me. So I started Navdanya, a movement that promotes biodiversity conservation and seed saving and seed sharing among farmers. Navdanya has created more than twenty community "seed banks" through which seeds are saved and freely exchanged among our three hundred thousand members.

Through our saving of heritage seeds, we have brought back "forgotten foods" like *jhangora* (barnyard millet), *ragi* (finger millet), *marsha* (amaranth), *naurangi dal,* and *gahat dal.* Not only are these crops more nutritious than the globally traded commodities, but they are also more resource prudent, requiring only two hundred to three hundred millimeters of rain compared to the twenty-five hundred millimeters needed for chemical rice farming. Millets could increase food production four hundred fold using the same amount of limited water. These forgotten foods are the foods of the future. Farmers' seeds are the seeds of the future.

The seed, for the farmer, is not merely the source of future plants/food; it is the storage place of culture, of history. Seed is the first link in the food chain. Seed is the ultimate symbol of food security.

Free exchange of seed among farmers has been the basis of maintaining biodiversity as well as food security. This exchange is based on cooperation and reciprocity. A farmer who wants to exchange seed generally gives an equal quantity of seed from his field in return for the seed he gets.

Free exchange among farmers goes beyond mere exchange of seeds; it involves the exchange of ideas and knowledge, of culture and heritage. It is an accumulation of tradition, of knowledge of how to work the seed. Farmers gather knowledge about the seeds they want to grow in future by watching them actually grow in other farmers' fields. This knowledge is based on the cultural, religious, and gastronomic values the community accords to the seed and the plant it produces as well as qualities of drought, disease, and pest resistance, longevity, and other aspects.

In saving seeds and biodiversity, we are protecting cultural diversity. Navdanya means "nine seeds." It also means "new gift." We bring to our farmers the new gift of life in the face of the extinction of species and the elimination of small farmers.

Our conservation of heritage rice varieties has led to the protection of the original, authentic basmati as part of the Slow Food Presidium. We have saved more than three thousand rice varieties, including over thirty aromatic rices. The saline-resistant seeds we have saved helped

Orissa farmers recover from the super cyclone that killed thirty thousand people in 1999. The saline-resistant seeds were also distributed by Navdanya for rehabilitation after the tsunami. We are now creating "Seeds of Hope" seed banks to deal with climate chaos. Heritage seeds that can tolerate droughts, floods, and cyclones will be collected, saved, multiplied, and distributed. Farmers' breeding is far ahead of scientific breeding and genetic engineering in providing flood-resistant, drought-resistant, saline-resistant varieties. In the context of farmers' heritage, genetic engineering is in fact a laggard technology.

Not only are corporate, industrial breeding strategies incapable of dealing with climate change, but genetically engineered seeds are also killing farmers. In India, according to a debate in Parliament, more than one hundred thousand farmers have committed suicide because of debt caused by the high cost of unreliable seeds sold by corporations. There are no suicides where farmers use heritage seeds and their own traditional varieties. Suicides are concentrated in areas that have become dependent on commercial seeds and are most intense where genetically engineered Bt cotton has been sold. These are seeds of suicide and seeds of slavery.

My inspiration for saving seeds came from Gandhi's spinning wheel, through which he fought the British Empire nonviolently. Another inspiration from Gandhi is the salt satyagraha, through which Gandhi refused to cooperate with salt laws that made salt a monopoly of the British. He walked to the sea, picked up salt, and said, "Salt making is our birthright." We have undertaken the "seed satyagraha"—a commitment to refuse to cooperate with patent laws and seed laws that prevent farmers from saving and exchanging seed. Seed freedom is our birthright. Without seed freedom there is no food freedom.

In May 2006, we in Navdanya undertook a "seed pilgrimage" (Bija Yatra) to stop farmers' suicides and create an agriculture of hope using heritage seeds and farmers' agroecological knowledge. The Bija Yatra was launched on May 10 to mark 150 years of our struggle for freedom. We are building a movement to stop the genocide of our farmers and reclaim our seed sovereignty and food sovereignty. The yatra started from Gandhi's ashram in Sevagram, District Wardha, Maharashtra, and concluded on May 26 in Bangalore. The yatra covered Amravati, Yavatmal, Nagpur in the Vidarbha region of Maharashtra, Adilabad,

Warrangal, Karimnagar, Hyderabad in Andhra Pradesh and Bidar, Gulbarga, Raichur, Hosepet, Chitradurg, and Bangalore in Karnataka. These are the regions where farmers have become locked into dependence on corporate seed supplies for growing cash crops integrated to world markets, which is leading to a collapse in farm prices due to $400 billion subsidies in rich countries.

Throughout the yatra seeds of freedom and seeds of life were distributed. We will boycott Monsanto's Bt cotton and poisonous agrochemicals that are killing our farmers and the environment. India is not free if its farmers are enslaved and indebted. We will not rest till our villages are GMO free, patent free, debt free, and suicide free.

Privatization of the earth's resources—of water, of biodiversity—is the ultimate social and ecological violation of the earth's rights and of human rights. The earth gives us gifts to be shared, to be conserved, to be used sustainably, to be returned to it in gratitude. The very idea of owning life through patents and owning and selling water through concessions and commodification is a symptom of the deep regression of the human species. So over the years resisting the enclosures of the commons and advocating for their recovery have defined my thoughts and my actions in my books *Biopiracy, Water Wars, Stolen Harvest,* and *Soil Not Oil.*

This is why I fought against the biopiracy patents on neem, basmati, and wheat. That is why I fought against the commodification of the Ganga and the privatization of Delhi's water supply, a story we have told in our reports *Ganga: Common Heritage or Commodity?* and *Water Democracy vs. Water Privatisation* in Delhi. Defending our fundamental freedoms has become fighting "free trade" to protect our seed freedom (*bija swaraj*), food freedom (*anna swaraj*), water freedom (*jal swaraj*), land freedom (*bhu swaraj*), and forest freedom. We have reinvented democracy as earth democracy, the democracy of all life and the democracy of everyday life (see my book *Earth Democracy*).

Most of the essays in this reader published by the University Press of Kentucky are extracted from the many books I have written over the years. The connecting thread is food and agriculture, a theme that has become a passion as deep as my passion for quantum theory was. What quantum theory has taught us about the false assumptions of the mechanistic worldview also holds true for a mechanistic view of farming as industrial agriculture. I did my doctoral thesis on nonseparability and nonlocality in quantum theory. Nonseparability and interconnectedness are even more true of the living world of seed and soil, plants and animals. Quantum theory teaches us that there are no fixed entities and fixed quantities in the world. The world in constant, dynamic interaction is a world of unfolding potential. There are no "essentially high-yielding varieties" and "low-yielding varieties." There are high-response varieties bred for chemical inputs, and there are farmers' varieties, which can give us high nutrients at low cost. The Navdanya report *Health per Acre* shows that native seeds and biodiverse agroecological systems produce more food and nutrition per acre than chemical monocultures.

The future of farming is based on love and care for the living soil, the living seed, and living food, all of which protect the well-being of the planet and its people.

1

The Gendered Politics of Food

Let them come and see men and women and children who know how
to live, whose joy of life has not yet been killed by those who claimed
to teach other nations how to live.

—Chinua Achebe

The Age of Enlightenment, and the theory of progress to which it gave
rise, was centered on the sacredness of two categories: modern scientific knowledge and economic development. Somewhere along the
way, the unbridled pursuit of progress, guided by science and development, began to destroy life without any assessment of how much of
the diversity of life on this planet is disappearing—and how fast. The
act of living, of celebrating and conserving life in all its diversity—in
people and in nature—seems to have been sacrificed to progress, and
the sanctity of life has been substituted by the sanctity of science and
development.

Throughout the world, a new questioning is growing, rooted in the
experience of those for whom the spread of what was called "enlightenment" has been the spread of darkness, of the extinction of life and
life-enhancing processes. A new awareness is growing that is questioning the sanctity of science and development and revealing that these are
not universal categories of progress but the special projects of modern
Western patriarchy. I have been involved with women's struggles for
survival in India over the last decade. My work is informed by both the
suffering and the insights of those who struggle to sustain and conserve
life, and whose struggles question the meaning of a progress, a science,
a development that destroys life and threatens survival.

The death of nature is central to this threat to survival. The earth is rapidly dying: its forests are dying, its soils are dying, its waters are dying, its air is dying. Tropical forests, the creators of the world's climate, the cradle of the world's vegetational wealth, are being bulldozed, burned, ruined, or submerged. In 1950, just over 100 million hectares of forests had been cleared—by 1975, this figure had more than doubled. During 1950–1975, at least 120 million hectares of tropical forests were destroyed in South and Southeast Asia alone; by the end of the century, another 270 million could be eliminated. In Central America and Amazonia, cattle ranching for beef production is claiming at least 2.5 million hectares of forests each year; in India 1.3 million hectares of forests are lost every year to commercial plantation crops, river valley projects, mining projects, and so on. Each year, 12 million hectares of forests are being eliminated from the face of the earth. At current rates of destruction, by the year 2050 all tropical forests will have disappeared, and with tropical forests will disappear the diversity of life they support.

Up to 50 percent of all living things—at least 5 million species— are estimated to live in tropical forests. A typical four-square-mile patch of rainforest contains up to 1,500 species of flowering plants, 750 species of trees, 125 of mammals, 400 of birds, 100 of reptiles, 60 of amphibians, and 150 of butterflies. The unparalleled diversity of species within tropical forests means relatively few individuals of each; any forest clearance thus disrupts their life cycles and threatens them with rapid extinction. Current estimates suggest that we are losing one species of life a day from the 5–10 million species believed to exist. If present trends continue, we can expect an annual rate of loss as high as 50,000 species by the year 2000. In India alone, there exist 7,000 species of plant life not found anywhere else in the world; the destruction of the country's natural forests implies the disappearance of this rich diversity of animal and plant life.

Forests are the matrix of rivers and water sources, and their destruction in tropical regions amounts to the desiccation and desertification of land. Every year 12 million hectares of land deteriorate into deserts and are unable to support vegetation or produce food. Sometimes land is laid waste through desertification, at other times through ill-conceived land use, which destroys the fertility of fragile tropical soils. Desertification in the Sahel in Africa has already killed millions of people and animals. Globally, some 456 million people today are starving or malnourished because of the desertification of croplands. Most agricultural lands cropped intensively with Green Revolution techniques are either waterlogged or desiccated deserts. Nearly 7 million hectares of land in India brought under irrigation have already gone out of production due to severe salinity, and an additional 6 million hectares have been seriously affected by water-logging. Green Revolution agriculture has decreased genetic diversity and increased the vulnerability of crops to failure through lowering resistance to drought and pests.

With the destruction of forests, water, and land, we are losing our life-support systems. This destruction is taking place in the name of "development" and progress, but there must be something seriously wrong with a concept of progress that threatens survival itself. The violence to nature, which seems intrinsic to the dominant development model, is also associated with violence to women who depend on nature for drawing sustenance for themselves, their families, their societies. This violence against nature and women is built into the very mode of perceiving both and forms the basis of the current development paradigm. My work is an attempt to articulate how rural Indian women, who are still embedded in nature, experience and perceive ecological destruction and its causes, and how they have conceived and initiated processes to arrest the destruction of nature and begin its regeneration. From the diverse and specific grounds of the experience of ecological destruction arise a common identification of its causes in the developmental process and the view of nature with which it is legitimized. I focus on science and development as patriarchal projects, not as a denial of other sources of patriarchy, such as religion, but because they are thought to be class, culture, and gender neutral.

Seen from the experiences of third world women, the modes of thinking and action that pass for science and development are not universal and humanly inclusive, as they are made out to be; modern science and development are projects of male, Western origin, both historically and ideologically. They are the latest and most brutal expression of a patriarchal ideology that is threatening to annihilate nature and the entire human species. The rise of a patriarchal science of nature took place in Europe during the same period as the closely related industrial revolution, which laid the foundations of a patriarchal mode of economic development in industrial capitalism. Contemporary science and development conserve the ideological roots and biases of the scientific and industrial revolutions even as they unfold into new areas of activity and new domains of subjugation.

The scientific revolution in Europe transformed nature from *terra mater* into a machine and a source of raw material; with this transformation it removed all ethical and cognitive constraints against its violation and exploitation. The industrial revolution converted economics from the prudent management of resources for sustenance and basic needs' satisfaction into a process of commodity production for profit maximization. Industrialism created a limitless appetite for resource exploitation, and modern science provided the ethical and cognitive license to make such exploitation possible, acceptable—and desirable. The new relationship of man's domination and mastery over nature was thus also associated with new patterns of domination and mastery over women, and their exclusion from participation *as partners* in both science and development.

Contemporary development activity in the third world superimposes the scientific and economic paradigms created by Western, gender-based ideology on communities in other cultures. Ecological destruction and the marginalization of women, we know now, have been the inevitable results of most development programs and projects based on such paradigms; they violate the integrity of one and destroy the productivity of the other. Women, as victims of the violence of patriarchal forms of development, have risen against them to protect nature and preserve their survival and sustenance. Indian women have been in the forefront of ecological struggles to conserve forests, land, and water. They have challenged the Western concept of nature as an object of exploitation and have protected it as Prakriti, the living force that supports life. They have challenged the Western concept of economics as production of profits and capital accumulation with their own concept of economics as production of sustenance and needs satisfaction. A science that does not respect nature's needs and a development that does not respect people's needs inevitably threaten survival. In their fight to survive the onslaughts of both, women have begun a struggle that challenges the most fundamental categories of Western patriarchy—its concepts of nature and women and of science and development. Their ecological struggle in India is aimed simultaneously at liberating nature from ceaseless exploitation and themselves from limitless marginalization. They are creating a feminist ideology that transcends gender, and a political practice that is humanly inclusive; they are challenging patriarchy's ideological claim to universalism—not with another universalizing tendency but with diversity; and they are challenging the dominant concept of power as violence with the alternative concept of nonviolence as power.

The everyday struggles of women for the protection of nature take place in the cognitive and ethical context of the categories of the ancient Indian worldview in which nature is Prakriti, a living and creative process, the feminine principle from which all life arises. Women's ecology movements, as the preservation and recovery of the feminine principle, arise from a nongender-based ideology of liberation, different both from the gender-based ideology of patriarchy that underlies the process of ecological destruction and women's subjugation and the gender-based responses that have, until recently, been characteristic of the West.

Inspired by women's struggles for the protection of nature as a condition for human survival, my work goes beyond a statement of women as special victims of the environmental crisis. It attempts to capture and reconstruct those insights and visions that Indian women provide in their struggles for survival that perceive development and science from outside the categories of modern Western patriarchy. These oppositional categories are simultaneously ecological and feminist: they allow the possibility of survival by exposing the parochial basis of science and development and by showing how ecological destruction and the marginalization of women are not inevitable, economically or scientifically.

Women of the third world have conserved those categories of thought and action that make survival possible, and therefore make justice and peace possible. Ecology movements, women's movements, and peace movements across the world can draw inspiration from these categories as forces of opposition and challenge to the dominant categories of Western patriarchy that rule the world today in the name of development and progress, even while they destroy nature and threaten the life of entire cultures and communities. I pay tribute to the leadership of millions of unknown women in India, struggling for a life that is simultaneously peaceful and just.

Note

The epigraph to this chapter is drawn from Chinua Achebe, *No Longer at Ease* (London: Heinemann, 1960), 45.

2

Science and Politics
in the Green Revolution

In 1970, Norman Borlaug was awarded the Nobel Peace Prize for "a new world situation with regard to nutrition." According to the Nobel Prize Committee, "The kinds of grain which are the result of Dr. Borlaug's work speed economic growth in general in the developing countries."[1] The "miracle seeds" that Borlaug had created were seen as a source of new abundance and peace. Science was applauded for having a magical ability to solve problems of material scarcity and violence.

"Green Revolution" is the name given to this science-based transformation of third world agriculture, and the Indian Punjab was its most celebrated success. Paradoxically, after two decades of the Green Revolution, Punjab is neither a land of prosperity nor of peace. It is a region riddled with discontent and violence. Instead of abundance, Punjab has been left with diseased soils, pest-infested crops, water-logged deserts, and indebted and discontented farmers. Instead of peace, Punjab has inherited conflict and violence. At least 15,000 people have lost their lives in the last six years. 598 people were killed in violent conflict in Punjab during 1986. In 1987 the number was 1,544. In 1988, it had escalated to 3,000. And 1989 shows no sign of peace in Punjab.

The tragedy of Punjab—of the thousands of innocent victims of violence over the past five years—has commonly been presented as an outcome of ethnic and communal conflict between two religious groups. This study presents a different aspect and interpretation of the Punjab tragedy. It introduces dimensions that have been neglected or gone unnoticed in understanding the emergent conflicts. It traces aspects of the conflicts and violence in contemporary Punjab to the ecological and political demands of the Green Revolution as a scientific experiment in development and agricultural transformation. The

15

Green Revolution has been heralded as a political and technological achievement, unprecedented in human history. It was designed as a strategy for peace through the creation of abundance by breaking out of nature's limits and variabilities. In its very genesis, the science of the Green Revolution was put forward as a political project for creating a social order based on peace and stability. However, when violence was the outcome of social engineering, the domain of science was artificially insulated from the domain of politics and social processes. The science of the Green Revolution was offered as a "miracle" recipe for prosperity. But when discontent and new scarcities emerged, science was delinked from economic processes.

On the one hand, contemporary society perceives itself as a science-based civilization, with science providing both the logic and the propulsion for social transformation. In this aspect science is self-consciously embedded in society. On the other hand, unlike all other forms of social organization and social production, science is placed above society. It cannot be judged, it cannot be questioned, it cannot be evaluated in the public domain. As Harding has observed, "Neither God nor tradition is privileged with the same credibility as scientific rationality in modern cultures. . . . The project that science's sacredness makes taboo is the examination of science in just the ways any other institution or set of social practices can be examined."[2]

While science itself is a product of social forces and has a social agenda determined by those who can mobilize scientific production, in contemporary times scientific activity has been assigned a privileged epistemological position of being socially and politically neutral. Thus science takes on a dual character. It offers technological fixes for social and political problems, but delinks itself from the new social and political problems it creates. Reflecting the priorities and perceptions of particular class, gender, or cultural interests, scientific thought organizes and transforms the natural and social order. However, since both nature and society have their own organization, the superimposition of a new order does not necessarily take place perfectly and smoothly. There is often resistance from people and nature, a resistance that is externalized as "unanticipated side effects." Science stays immune from social assessment, and insulated from its own impacts. Through this split identity is created the "sacredness" of science.

Within the structure of modern science itself are characteristics that prevent the perception of linkages. Fragmented into narrow disciplines and reductionist categories, scientific knowledge has a blind spot with respect to relational properties and relational impacts. It tends to decontextualize its own context. Through the process of decontextualization, the negative and destructive impacts of science on nature and society are externalized and rendered invisible. Being separated from their material and political roots in the science system, new forms of scarcity and social conflict are then linked to other social systems, for example, religion.

The conventional model of science, technology, and society locates sources of violence in politics and ethics, in the application of science and technology, not in scientific knowledge itself.[3] The assumed dichotomy between values and facts underlying this model implies a dichotomy between the world of values and the world of facts. In this view, sources of violence are located in the world of values, while scientific knowledge inhabits the world of facts.

The fact-value dichotomy is a creation of modern reductionist science which, while being an epistemic response to a particular set of values, posits itself as independent of values. By splitting the world into fact versus values, it conceals the real difference between two kinds of value-laden facts. Modern reductionist science is characterized in the received view as the discovery of the properties and laws of nature in accordance with a "scientific" method that claims to be "objective," "neutral," and "universal." This view of reductionist science as being a description of reality as it is, unprejudiced by value, is being rejected increasingly on historical and philosophical grounds. It has been historically established that all knowledge, including modern scientific knowledge, is built on the use of a plurality of methodologies, and reductionism itself is only one of the scientific options available.

The knowledge and power nexus is inherent to the reductionist system because the mechanistic order, as a conceptual framework, was associated with a set of values based on power that was compatible with the needs of commercial capitalism. It generates inequalities and domination by the way knowledge is generated and structured, the way it is legitimized, and by the way in which such knowledge transforms nature and society.

The experience of the Green Revolution in Punjab is an illustration of how contemporary scientific enterprise is politically and socially created, how it builds its immunity and blocks its social evaluation. It is an example of how science takes credit for successes and absolves itself from all responsibility for failures. The tragic story of Punjab is a tale of the exaggerated sense of modern science's power to control nature and society, and the total absence of a sense of responsibility for creating natural and social situations that are totally out of control. The externalization of the consequences of the Green Revolution from the scientific and technological package of the Green Revolution has been, in our view, a significant reason for the communalization of the Punjab crises.

It is, however, misleading to reduce the roots of the Punjab crisis to religion, as most scholars and commentators have done, since the conflicts are also rooted in the ecological, economic, and political impacts of the Green Revolution. They are not merely conflicts between two religious communities but reflect tensions between a disillusioned and discontented farming community and a centralizing state that controls agricultural policy, finance, credit, and inputs and prices of agricultural commodities. At the heart of these conflicts and disillusionments lies the Green Revolution.

This essay presents the other side of the Green Revolution story— its social and ecological costs hidden and hitherto unnoticed. In so doing, it also offers a different perspective on the multiple roots of ethnic and political violence. It illustrates that ecological and ethnic fragmentation and breakdown are intimately connected and are an intrinsic part of a policy of planned destruction of diversity in nature and culture to create the uniformity demanded by centralized management systems. The ecological and ethnic crises in Punjab can be viewed as arising from a basic and unresolved conflict between the demands of diversity, decentralization, and democracy on the one hand, and the demands of uniformity, centralization, and militarization on the other. Control over nature and control over people were essential elements of the centralized and centralizing strategy of the Green Revolution. Ecological breakdown in nature and the political breakdown of society were consequences of a policy based on tearing apart both nature and society.

The Green Revolution was based on the assumption that technology is a superior substitute for nature, and hence a means of producing limitless growth, unconstrained by nature's limits. However, the assumption of nature as a source of scarcity, and technology as a source of abundance, leads to the creation of technologies that produce new scarcities in nature through ecological destruction. The reduction in the availability of fertile land and genetic diversity of crops as a result of Green Revolution practices indicates that at the ecological level, the Green Revolution produced scarcity, not abundance.

Not just ecological insecurity but also social and political insecurity was generated by the Green Revolution. Instead of stabilizing and pacifying the countryside, it fueled a new pattern of conflict and violence. The communalization of the Punjab conflicts that originally arose from the processes of political transformation associated with the Green Revolution was based, in part, on externalizing the political impacts of technological change from the domain of science and technology. A similar pattern of externalization seems to be at play in the introduction of the "biotechnology revolution," exemplified in Punjab by the Pepsi project.

The social and political planning that went into the Green Revolution aimed at engineering not just seeds but social relations as well. Punjab is an exemplar of how this engineering went out of control both at the material and the political level.

The Green Revolution and the Conquest of Nature

Half a century ago, Sir Albert Howard, the father of modem sustainable farming, wrote in his classic *An Agricultural Testament*, "In the agriculture of Asia we find ourselves confronted with a system of peasant farming which, in essentials, soon became stabilized. What is happening today in the small fields of India and China took place many centuries ago. The agricultural practices of the orient have passed the supreme test, they are almost as permanent as those of the primeval forest, of the prairie, or of the ocean."[4]

In 1889, Dr. John Augustus Voelcker was deputed by the secretary of state to India to advise the imperial government on the application of agricultural chemistry to Indian agriculture. In his report to the Royal

Agricultural Society of England on the improvement of Indian Agriculture, Voelcker stated:

> I explain that I do not share the opinions which have been expressed as to Indian Agriculture being, as a whole, primitive and backward, but I believe that in many parts *there is little or nothing that can be improved.... Where agriculture is manifestly inferior, it is more generally the result of the absence of facilities which exist in the better districts than from inherent bad systems of cultivation. . . .* I may be bold to say that it is a much easier task to propose improvements in English agriculture than to make really valuable suggestions for that of India. To take the ordinary acts of husbandry, no where would one find better instances of keeping land scrupulously clean from weeds, of ingenuity in device of water raising appliances, of knowledge of soils and their capabilities as well as of the exact time to sow and to reap as one would in Indian agriculture, and this not at its best only but at its ordinary level. It is wonderful, too, how much is known of rotation, the system of mixed crops and of fallowing. Certain it is that I, at least, have never seen a more perfect picture of careful cultivation combined with hard labour, perseverance and fertility of resource.[5]

When the best of Western scientists were earlier sent to "improve" Indian agriculture, they found nothing that could be improved in the principles of farming, which were based on preserving and building on nature's process and nature's patterns. Where Indian agriculture was less productive, it was due not to primitive principles or inferior practices but to interruptions in the flow of resources that made productivity possible. Land alienation, the reservation of forests, and the expansion of cash crop cultivation were among the many factors, introduced during colonialism, that created a scarcity of local inputs of water and manure to maintain agricultural productivity.

In the second quarter of the century, from World War I to independence, Indian agriculture suffered a setback as a consequence of complex factors including reduced exports due to worldwide recession, depression, and the near-complete paralysis of shipping during World

War II. The chaos of partition added to its decline, and the expansion of commercial crops like sugarcane and groundnuts pushed food grains onto poorer lands where yields per acre were lower. The upheavals during this period left India faced with a severe food crisis.

There were two responses to the food crisis created through the war years and during partition. The first was indigenous, the second exogenous. The indigenous response was rooted in the independence movement. It aimed at strengthening the ecological base of agriculture and the self-reliance of the peasants of the country. The *Harijan,* a newspaper published by Mahatma Gandhi that had been banned from 1942 to 1946, was full of articles written by Gandhi during 1946–1947 on how to deal with food scarcity politically, and by Mira Behn, Kumarappa, and Pyarelal on how to grow more food using internal resources. On June 10, 1947, referring to the food problem at a prayer meeting, Gandhi said:

> The first lesson we must learn is of self-help and self-reliance. If we assimilate this lesson, we shall at once free ourselves from disastrous dependence upon foreign countries and ultimate bankruptcy. This is not said in arrogance but as a matter of fact. We are not a small place, dependent for this food supply upon outside help. We are a sub-continent, a nation of nearly 400 millions. We are a country of mighty rivers and a rich variety of agricultural land, with inexhaustible cattle-wealth. That our cattle give much less milk than we need, is entirely our own fault. Our cattle-wealth is any day capable of giving us all the milk we need. Our country, if it had not been neglected during the past few centuries, should not today only be providing herself with sufficient food, but also be playing a useful role in supplying the outside world with much-needed foodstuffs of which the late war has unfortunately left practically the whole world in want. This does not exclude India.[6]

Recognizing that the crisis in agriculture was related to a breakdown of nature's processes, India's first agriculture minister, K. M. Munshi, had worked out a detailed strategy on rebuilding and regenerating the ecological base of productivity in agriculture based on a bottom-up, decentralized, and participatory methodology. In a seminar on September 27, 1951, organized by the Agriculture Ministry, a program of regeneration of Indian agriculture was worked out, with the recognition that the diversity of India's soils, crops, and climates had to be taken into account. The need to plan from the bottom, to consider every individual village and sometimes every individual field, was considered essential for the program, called "land transformation." At this seminar, K. M. Munshi told the state directors of agricultural extension:

> Study the Life's Cycle in the village under your charge in both its aspects—hydrological and nutritional. Find out where the cycle has been disturbed and estimate the steps necessary for restoring it. Work out the village in four of its aspects, (1) existing conditions, (2) steps necessary for completing the hydrological cycle, (3) steps necessary to complete the nutritional cycle, and a complete picture of the village when the cycle is restored, and (4) have faith in yourself and the programme. Nothing is too mean and nothing too difficult for the man who believes that the restoration of the life's cycle is not only essential for freedom and happiness of India but is essential for her very existence.[7]

Repairing nature's cycles and working in partnership with nature's processes were viewed as central to the indigenous agricultural policy.

However, while Indian scientists and policy makers were working out self-reliant and ecological alternatives for the regeneration of agriculture in India, another vision of agricultural development was taking shape in American foundations and aid agencies. This vision was based not on cooperation with nature but on its conquest. It was based not on the intensification of nature's processes but on the intensification of credit and purchased inputs like chemical fertilizers and pesticides. It was based not on self-reliance but on dependence. It was based not on diversity but uniformity. Advisors and experts came from America to shift India's agricultural research and agricultural policy from an indigenous and ecological model to an exogenous and high-input one, finding, of course, partners in sections of the elite, because the new model suited their political priorities and interests.

There were three groups of international agencies involved in transferring the American model of agriculture to India—private American foundations, the American government, and the World Bank. The Ford Foundation had been involved in training and agricultural extension since 1952. The Rockefeller Foundation had been involved in remodeling the agricultural research system in India since 1953. In 1958, the Indian Agricultural Research Institute, which had been set up in 1905, was reorganized, and Ralph Cummings, the field director of the Rockefeller Foundation, became its first dean. In 1960, he was succeeded by A. B. Joshi, and in 1965 by M. S. Swaminathan. Besides reorganizing Indian research institutes on American lines, the Rockefeller Foundation also financed the trips of Indians to American institutions. Between 1956 and 1970, ninety short-term travel grants were awarded to Indian leaders to see American agricultural institutes and experimental stations. One hundred and fifteen trainees finished studies under the foundation. Another two thousand Indians were financed by USAID to visit the United States for agricultural education during the period. The work of the Rockefeller and Ford foundations was facilitated by agencies like the World Bank, which provided the credit to introduce a capital-intensive agricultural model in a poor country. In the mid-1960s, India was forced to devalue its currency to the extent of 37.5 percent. The World Bank and USAID also exerted pressure for favorable conditions for foreign investment in India's fertilizer industry, import liberalization, and elimination of domestic controls. The World Bank provided credit for the foreign exchange needed to implement these policies. The foreign exchange component of the Green Revolution strategy over the five-year plan period (1966–1971) was projected to be Rs. 1,114 crores, which converted to about $2.8 billion at the then official rate. This was a little over six times the total amount allocated to agriculture during the preceding third plan (Rs. 191 crores). Most of the foreign exchange was needed for the import of fertilizers, seeds, and pesticides, the new inputs in a chemically intensive strategy. The World Bank and USAID stepped in to provide the financial input for a technology package that the Ford and Rockefeller foundations had evolved and transferred.

Within India, the main supporter of the Green Revolution strategy was C. Subramaniam, who became agriculture minister in 1964, and M. S. Swaminathan, who became the director of the Indian Agricultural Research Institute (IARI) in 1965 and had been trained by Norman Borlaug, who worked for Rockefeller's agricultural program in Mexico. After a trip to India in 1963, he dispatched four hundred kilograms of semidwarf varieties to be tested in India. In 1964, rice seeds were brought in from the International Rice Research Institute (IRRI) in the Philippines (which had recently been set up with Ford and Rockefeller funds). In the same year Ralph Cummings felt that sufficient testing had been done to release the varieties on a large scale. He approached C. Subramaniam to see if the new agriculture minister would be willing to throw his support to accelerating the process of introducing the Green Revolution seeds. Subramaniam acknowledges that he decided to follow Cummings's advice quickly and began to formulate a strategy for using the new varieties.[8]

Others in India were not as willing to adopt the American agricultural strategy. The Planning Commission was concerned about the foreign-exchange costs of importing the fertilizer needed for application to the high-yielding varieties (HYVs) in a period of a severe balance-of-payments crisis. Leading economists B. S. Minhas and T. S. Srinivas questioned the strategy on economic grounds. State governments worried that adoption of the new seeds would reduce their autonomy in agricultural research. Agricultural scientists objected to the new varieties because of the risks of disease and the displacement of small peasants. The only ones supporting Subramaniam were the younger agricultural scientists trained over the past decade in the American paradigm of agriculture.

The occurrence of drought in 1966 caused a severe drop in food production in India and an unprecedented increase in food grain supply from the United States. Food dependency was used to set new policy conditions on India. The U.S. president, Lyndon Johnson, put wheat supplies on a short tether. He refused to commit food aid beyond one month in advance until an agreement to adopt the Green Revolution package was signed between the Indian agriculture minister, C. Subramaniam, and the U.S. secretary of agriculture, Orville Freeman.[9]

Lal Bahadur Shastri, the Indian prime minister in 1965, had counseled caution against rushing into a new agriculture based on new varieties. With his sudden death in 1966 the new strategy was easier to

introduce. The Planning Commission, which approves all large investment in India, was also bypassed since it was viewed as a bottleneck.

Rockefeller agricultural scientists saw third world farmers and scientists as not having the ability to improve their own agriculture. They
believed that the answer to greater productivity lay in the American-
style agricultural system. However, the imposition of the American
model of agriculture did not go unchallenged in the third world or in
America. Edmundo Taboada, who was head of the Mexican office of
Experiment Stations, maintained, like K. M. Munshi in India, that ecologically and socially appropriate research strategies could evolve only
with the active participation of the peasantry. "Scientific Research must
take into account the men that will apply its results. . . . Perhaps a discovery may be made in the laboratory, a greenhouse or an experimental station, but useful science, a science that can be applied and handled
must emerge from the local laboratories of . . . small farmers, ejidatorios
and local communities."[10]

Together, peasants and scientists searched for ways to improve the
quality of *criollo* seeds (open-pollinated indigenous varieties) that could
be reproduced in peasant fields. However, by 1945, the Special Studies
Bureau in the Mexican Agriculture Ministry, funded and administered
by the Rockefeller Foundation, had eclipsed the indigenous research
strategy and started to export to Mexico the American agricultural revolution. In 1961, the Rockefeller-financed center took the name of CIM-
MYT (Centro international de mejoramiento de maiz y trigo, or the
International Maize and Wheat Improvement Center). The American
strategy, reinvented in Mexico, then came to the entire third world as
the "Green Revolution."

The American model of agriculture had not done too well in America, though its nonsustainability and high ecological costs went ignored.
The intensive use of artificial fertilizers, extensive practice of monocultures, and intensive and extensive mechanization had turned fertile
tracts of the American prairies into a desert in less than thirty years.
The American Dust Bowl of the 1930s was in large measure a creation
of the American agricultural revolution. Hyams reports,

> When, between 1889 and 1900, thousands of farmers were
> settling in Oklahoma, it must have seemed to them that they

were founding a new agricultural civilization which might endure as long as Egypt. The grandsons, and even the sons of these settlers who so swiftly became a disease of their soil, trekked from their ruined farmsteads, their buried or uprooted crops, their dead soil, with the dust of their own making in their eyes and hair, the barren sand of a once fertile plain gritting between their teeth. The pitiful procession passed westward, an object of disgust—the God-dam'd Okies. But these God-dam'd Okies were the scapegoats of a generation, and the God who had damned them was perhaps after all a Goddess, her name Ceres, Demeter, Maia, or something older and more terrible. And what she damned them for was their corruption, their fundamental ignorance of the nature of her world, their defiance of the laws of co-operation and return which are the basis of life on this planet.[11]

When an attempt was made to spread this ecologically devastating vision of agriculture to other parts of the world through Rockefeller Foundation programs, notes of caution were sounded.

The American strategy of the Rockefeller and Ford foundations differed from the indigenous strategies primarily in the lack of respect for nature's processes and people's knowledge. In mistakenly identifying the sustainable and lasting as backward and primitive, and in perceiving nature's limits as constraints on productivity that had to be removed, American experts spread ecologically destructive and unsustainable agricultural practices worldwide. The Ford Foundation had been involved in agricultural development in India since 1951. In 1952, fifteen community development projects, each covering about one hundred villages, were started with Ford Foundation financial assistance. This program was, however, shed in 1959 when a Ford Foundation mission of thirteen North American agronomists to India argued that it was impossible to make simultaneous headway in all of India's 550,000 villages. Their recommendations for a selective and intensive approach among farmers and among districts led to the winding down of the community development program and the launching of the Intensive Agricultural Development Programme (IADP) in 1960–1961.

The IADP totally replaced an indigenous, bottom-up, organic-based strategy for regenerating Indian agriculture with an exogenous, top-down, chemically intensive one. Industrial inputs like chemical fertilizers and pesticides were seen as breaking Indian agriculture out of the "shackles of the past," as an article, "The Foundations Involvement in Intensive Agricultural Development in India," stated:

> India is richly endowed with sunshine, vast land areas (much of it with soils responsive to modernizing farming), a long growing season (365 days a year in most areas). Yet the solar energy, soil resources, crop growing days and water for irrigation are seriously underused or misused. India's soils and climate are among the most underused in the world. Can multiple cropping help Indian farmers utilize these vast resources more effectively—the answer must be yes.
>
> New opportunities for intensifying agricultural programs through multiple cropping are presenting themselves; led by the plant breeder there are new short season, fertilizer responsive, non-photo sensitive crops and varieties that under skillful farming practices have high yield potential; chemical fertilizer supplies are increasing rapidly—this frees the Indian cultivator from the shackles of the past permitting only very modest improvement of soil fertility through green manure and compost and the slow, natural recharge of soil nutrients. Also, up until recently varieties were bred for these conditions, plant protection was applied after the damage was done, and so on—a status quo agriculture. This has changed. Indian farmers are prepared to innovate and change; Indian leaders in agricultural development, extension, research and administration are beginning to understand the new potentials; intensive agriculture, first identified under IADP, is now India's food production strategy.[12]

Under the Ford Foundation program, agriculture was transformed from one based on internal inputs that are easily available at no cost to one dependent on external inputs for which credits became necessary. Instead of promoting the importance of agriculture in all regions, the IADP showed favoritism to specially selected areas for agricultural development, to which material and financial resources of the entire country were diverted. The latter, however, was a failed strategy where native varieties of food crops were concerned. The native crops tend to "lodge," or fall, under the intensive application of chemical fertilizers, thus putting a limit to fertilizer use. As a spokesperson of the Ford Foundation put it, "The programme revealed the urgent need for improved crop varieties as it was found that the native varieties (the only ones available during these early years) responded very poorly to improved practices and produced low yields even when subjected to other modern recommended practices."

It was not that native crop varieties were low yielding inherently. The problem with indigenous seeds was that they could not be used to consume high doses of chemicals. The Green Revolution seeds were designed to overcome the limits placed on chemically intensive agriculture by the indigenous seeds. The new seeds thus became central to breaking out of nature's limits and cycles. The "miracle" seeds were therefore at the heart of the science of the "Green Revolution."

The combination of science and politics in creating the Green Revolution goes back to the period in the 1940s when Daniels, the U.S. ambassador to the government of Mexico, and Henry Wallace, vice president of the United States, set up a scientific mission to assist in the development of agricultural technology in Mexico. The Office of Special Studies was set up in Mexico in 1943 within the Agricultural Ministry as a cooperative venture between the Rockefeller Foundation and the Mexican government. In 1944, Dr. J. George Harrar, head of the new Mexican research program, and Dr. Frank Hanson, an official of the Rockefeller Foundation in New York, invited Norman Borlaug to shift from his classified wartime laboratory job in Dupont to the plant-breeding program in Mexico. By 1954, Borlaug's "miracle seeds" of dwarf varieties of wheat had been bred. In 1970, Borlaug had been awarded the Nobel Peace Prize for his "great contribution towards creating a new world situation with regard to nutrition. . . . The kinds of grain which are the result of Dr Borlaug's work speed economic growth in general in the developing countries."[13] This assumed link between the new seeds and abundance, and between abundance and peace, was sought with the goal of replicating it rapidly in other regions of the world, especially Asia.

Impressed with the successful diffusion of "miracle" seeds of wheat from CIMMYT, which had been set up on the basis of the Rockefeller Foundation and Mexican government program, the Rockefeller and Ford foundations in 1960 established IRRI, which by 1966 was producing "miracle" rice to join the "miracle" wheats from CIMMYT.

CIMMYT and IRRI were the international agricultural research centers that grew out of the Rockefeller Foundation country program to launch the new seeds and the new agriculture across Latin America and Asia. By 1969, the Rockefeller Foundation, in cooperation with the Ford Foundation, had established the Centro international de agriculture tropical (CIAT) in Colombia and the International Institute for Tropical Agriculture (IITA) in Nigeria.

In 1971, at the initiative of Robert McNamara, the president of the World Bank, the Consultative Group on International Agricultural Research (CGIAR) was formed to finance the network of these international agricultural centers (IARC). Later, nine more IARCs were added to the CGIAR system. The International Crops Research Institute for the Semi-Arid Tropics (ICRISAT) was started in Hyderabad in India in 1971. The International Laboratory for Research on Animal Diseases (ILRAD) and the International Livestock Centre for Africa (ILCA) were approved in 1973. The Consultative Group had sixteen donors, who contributed $20.06 million in 1972. By 1981, the budget had shot up to $157.945 million provided by forty donors.

The growth of the international institutes was based on the erosion of the decentralized knowledge systems of third world peasants and third world research institutes. The centralized control of knowledge and genetic resources was, as mentioned, not achieved without resistance. In Mexico, peasant unions protested against it. Students and professors at Mexico's National Agricultural College in Chapingo went on strike to demand a program different from the one that emerged from the American strategy and was more suitable to the small-scale poor farmers and to the diversity of Mexican agriculture.

The International Rice Research Institute was set up in 1960 by the Rockefeller and Ford foundations, nine years after the establishment of a premier Indian Institute, the Central Rice Research Institute (CRRI) in Cuttack. The Cuttack institute was working on rice research based on indigenous knowledge and genetic resources, a strategy clearly in conflict with the American-controlled strategy of the International Rice Research Institute. Under international pressure, the director of CRRI was removed when he resisted handing over his collection of rice germplasm to IRRI, and when he asked for restraint in the hurried introduction of the HYVs from IRRI.

The Madhya Pradesh government gave a small stipend to the ex-director of CRRI so that he could continue his work at the Madhya Pradesh Rice Research Institute (MPRRI) at Raipur. On this shoestring budget, he conserved twenty thousand indigenous rice varieties in situ in India's rice bowl in Chattisgarh. Later the MPRRI, which was doing pioneering work in developing a high-yielding strategy based on the indigenous knowledge of the Chattisgarh tribals, was also closed down due to pressure from the World Bank (which was linked to IRRI through CGIAR) because MPRRI had reservations about sending its collection of germplasm to IRRI.[14]

In the Philippines, IRRI seeds were called "Seeds of Imperialism." Burton Onate, president of the Philippines Agricultural Economics and Development Association, observed that IRRI practices had created debt and a new dependence on agrochemicals and seeds. "This is the Green Revolution connection," he remarked. "New seeds from the CGIAR global crop/seed systems which will depend on the fertilizers, agrichemicals and machineries produced by conglomerates of transnational corporations."[15]

Centralism of knowledge was built into the chain of CGIARs, from which technology was transferred to second-order national research centers. The diverse knowledge of local cultivators and plant breeders was displaced. Uniformity and vulnerability were built into international research centers run by American and American-trained experts breeding a small set of new varieties that would displace the thousands of locally cultivated plants in the agricultural systems, built up over generations on the basis on knowledge generated over centuries.

Politics was built into the Green Revolution because the technologies created were directed at capital-intensive inputs for best endowed farmers in the best endowed areas, and directed away from resource-prudent options of the small farmer in resource-scarce regions. The science and technology of the Green Revolution excluded poor regions and poor people as well as sustainable options. American advisors used the slogan of "building on the best." The science of the Green Revolution was thus essentially a political choice.

As Lappe and Collins have stated: "Historically, the Green Revolution represented a choice to breed seed varieties that produce high yields under optimum conditions. It was a choice not to start by developing seeds better able to withstand drought or pests. It was a choice not to concentrate first on improving traditional methods of increasing yields, such as mixed cropping. It was a choice not to develop technology that was productive, labour-intensive, and independent of foreign input supply. It was a choice not to concentrate on reinforcing the balanced, traditional diets of grain plus legumes."[16]

The crop and varietal diversity of indigenous agriculture was replaced by a narrow genetic base and monocultures. The focus was on internationally traded grains and a strategy of eliminating mixed and rotational cropping and replacing diverse varieties with varietal simplicity. While the new varieties reduced diversity, they increased resource use of water and boosted the employment of chemical inputs such as pesticides and fertilizers.

The strategy of the Green Revolution was aimed at transcending scarcity and creating abundance. Yet it put new demands on scarce renewable resources and generated new demands for nonrenewable resources. The Green Revolution technology required heavy investments in fertilizers, pesticides, seed, water, and energy. Intensive agriculture generated severe ecological destruction, created new kinds of scarcity and vulnerability, and resulted in new levels of inefficiency in resource use. Instead of transcending the limits imposed by natural endowments of land and water, the Green Revolution introduced new constraints on agriculture by wasting and destroying land, water resources, and crop diversity. The Green Revolution had been offered as a miracle, yet, as Angus Wright has observed: "One way in which agricultural research went wrong was precisely in saying and allowing it to be said that some miracle was being produced. . . . Historically, science and technology made their first advances by rejecting the idea of miracles in the natural world. Perhaps it would be best to return to that position."[17]

The Green Revolution and the Control of Society

The Green Revolution was promoted as a strategy that would simultaneously create material abundance in agricultural societies and reduce agrarian conflict. The new seeds of the Green Revolution were to be seeds of plenty and were also to be the seeds of a new political economy in Asia.

The Green Revolution was necessarily paradoxical. On the one hand, it offered technology as a substitute for both nature and politics in the creation of abundance and peace. On the other hand, the technology itself demanded more intensive natural resource use along with intensive external inputs and involved a restructuring of the way power was distributed in society. While treating nature and politics as dispensable elements in agricultural transformation, the Green Revolution created major changes in natural ecosystems and agrarian structures. New relationships between science and agriculture defined new links between the state and cultivators, between international interests and local communities, and within the agrarian society.

The Green Revolution was not the only strategy available. There was another strategy for agrarian peace based on reestablishing justice through land reform and the removal of political polarization, which was at the base of political unrest in agrarian societies.

Colonialism had dispossessed peasants throughout the third world of their entitlements to land and to full participation in agricultural production. In India, the British introduced the system of zamindari, or landlordism, to help divert land from growing food to growing opium and indigo as well as to extract revenue from the cultivators. R. P. Dutt records the sudden increase in agricultural revenues when the East India Company of British soldier-traders took over revenue rights of Bengal: "In the last year of administration of the last Indian ruler of Bengal, in 1764–65, the land revenue realized was £817,000. In the first year of the company's administration, in 1765–66, the land revenue realized in Bengal was £1,470,000. By 1771–72, it was £2,348,000 and by 1775–76, it was £2,818,000. When Lord Cornwallis fixed the permanent settlement in 1793, he fixed £3,400,000."[18]

The diversion of increasing amounts of agricultural produce as a source of colonial revenue took its toll in terms of deteriorating conditions of peasants and agricultural production. According to Bajaj:

> With more and more money flowing into the British hands the village and the producer were left with precious little to feed themselves and maintain the various village institutions that catered to their needs. According to Dharampal's estimates, whereas around 1750, for every 1,000 units of produce the producer paid 300 as revenue, only 50 of which went out to the central authority, the rest remaining within the village; by 1830,

he had to give away 650 units as revenue, 590 of which went straight to the central authority. As a result of this level of revenue collection the cultivators and the villagers both were destroyed.[19]

In Mexico, the Spanish instituted the system of *hacienda* (large estate) owners. After two centuries of colonization, haciendas dominated the countryside. They covered 70 million hectares of the land, leaving only 18 million hectares under the control of indigenous communities. According to Esteva, by 1910, around 8,000 haciendas were in the hands of a small number of owners, occupying 113 million hectares, with 4,500 managers, 300,000 tenants, and 3 million indentured peons and sharecroppers. An estimated 150,000 "Indian" communal landholders occupied 6 million hectares. Less than 1 percent of the population owned over 90 percent of the land, and over 90 percent of the rural population lacked any access to it.[20]

Between 1910 and 1917, over 1 million peasants in Mexico had died fighting for land. Between 1934 and 1949, Lazaro Cardenas redistributed 78 million acres, benefiting 42 percent of the entire agricultural population. Under the new distribution, small farmers owned 47 percent of the land. As Lappe and Collins report:

> Social and economic process was being achieved not through dependence on foreign expertise or costly imported agricultural inputs but rather with the abundant, underutilized resources of local peasants. While production increases were seen as important, the goal was to achieve them through helping every peasant to be productive, for only then would the rural majority benefit from the production increases. Freed from the fear of landlords, bosses, and money-lenders, peasants were motivated to produce, knowing that at last they would benefit from their own labor. Power was perceptibly shifting to agrarian reform organizations controlled by those who worked the fields.[21]

The result of this gain in political and economic power of the peasants was the erosion of power of the powerful hacienda owners and of the U.S. corporate sector, whose investment dropped by about 40 percent between the mid-1930s and the early 1940s.

When Cardenas was succeeded by Avila Camacho, a fundamental shift was induced in Mexico's agricultural policy. It was now to be guided by American control over research and resources for agriculture through the Green Revolution strategy. Peasant movements had tried to restructure agrarian relationships through the recovery of land rights. The Green Revolution tried to restructure social relationships by separating issues of agricultural production from issues of justice. Green Revolution politics was primarily a politics of depoliticization. According to Anderson and Morrison: "The founding of the International Rice Research Institute in Los Banos in 1960 was the institutional embodiment of the conviction that high quality agricultural research and its technological extensions would increase rice production, ease the food supply situation, spread commercial prosperity in the rural areas, and defuse agrarian radicalism."[22]

In the 1950s, the newly independent countries of Asia were faced with rising peasant unrest. When the Chinese Communist Party came to power, it had encouraged local peasants' associations to seize land, cancel debts, and redistribute wealth. Peasant movements inspired by the Chinese experience flared up in the Philippines, Indonesia, Malaysia, Vietnam, and India. The new political authorities in these Asian countries had to find a means to control agrarian unrest and stabilize the political situation. This "would include defusing the most explosive grievances of the more important elements in the countryside."[23]

In India land reforms had been viewed as a political necessity at the time of independence. Most states had initiated land reforms by 1950 in the form of abolition of Zamindari, security of tenure for tenant cultivators, and fixation of reasonable rents. Ceilings on landholdings were also introduced. In spite of weaknesses in the application of land-reform strategies, they provided relief to cultivators through the 1950s and 1960s. Aggregate crop output kept increasing during the 1950s in response to the restoration of some just order in land relations.

A second strategy for agricultural production and agrarian peace was, however, being worked out internationally, driven by concern at the "loss of China." American agencies like the Rockefeller and Ford foundations, U.S. Aid, the World Bank, and the like mobilized themselves for a new era of political intervention. As Anderson and Morrison have observed: "Running through all these measures, whether major or minor in their effect, was the concern to stabilize the countryside politically. It was recognised internationally that the peasantry were incipient revolutionaries and if squeezed too hard could be rallied against the new bourgeois-dominated governments in Asia. This recognition led many of the new Asian governments to join the British-American-sponsored Colombo Plan in 1952 which explicitly set out to improve conditions in rural Asia as a means of defusing the Communist appeal. Rural development assisted by foreign capital was prescribed as a means of stabilizing the countryside."[24] In Cleaver's view: "Food was clearly recognised as a political weapon in the efforts to thwart peasant revolution in many places in Asia . . . from its beginning the development of the Green Revolution grains constituted mobilizing science and technology in the service of counter-revolution."[25]

Science and politics were thus wedded together in the very inception of the Green Revolution as a strategy for increasing material prosperity and hence defusing agrarian unrest. For the social planners in national governments and international aid agencies, the science and technology of the Green Revolution were an integral part of sociopolitical strategy aimed at pacifying the rural areas of developing nations in Asia, not through redistributive justice but through economic growth. And agriculture was to be the source of this new growth.

While the Green Revolution was clearly political in reorganizing agricultural systems, concern for political issues such as participation and equity was consciously bypassed, replaced by the political concern for stability. Goals of growth had to be separated from goals of political participation. As David Hopper, then with Rockefeller Foundation, wrote in his "Strategy for the Conquest of Hunger":

> Let me begin my examination of the essentials for pay-off by focussing on public policy for agricultural growth. The confusion of goals that has characterized purposive activity for agricultural development in the past cannot persist if hunger is to be overcome. National governments must clearly separate the goal of growth from the goals of social development and political participation. . . . These goals are not necessarily incompatible, but their joint pursuit in unitary action programs is incompatible with development of an effective strategy for abundance. To conquer hunger is a large task. To ensure social equity and opportunity is another large task. Each aim must be held separately and pursued

by separate action. Where there are complementarities they should be exploited. But conflict in programme content must be solved quickly at the political level with a full recognition that if the pursuit of production is made subordinate to these aims, the dismal record of the past will not be altered.[26]

The record of the achievements of increased production through distributive justice is available in the experience of both Mexico and India in the years prior to the Green Revolution.

Gustavo Esteva reports that as a result of the land reforms of the 1930s, the *ejidos,* or lands, returned to peasant communities accounted for more than half of the total arable land of the country, and by 1940, for 51 percent of the total agricultural production. The production of the period continually expanded, at an annual rate of 5.2 percent from 1935 to 1942. Similarly, Jatindar Bajaj, in his study of pre– and post–Green Revolution performance, shows that the rate of growth of aggregate crop production was higher in the years before the Green Revolution than after it. The year 1967–1968 is when the Green Revolution was officially launched in India (see table 2.4).

The record of agricultural production before the Green Revolution was clearly not "dismal." Nor has the record of production been miraculous since the introduction of the "miracle" seeds. The usual argument created to support the image of the "miracle" is that India was transformed from "the begging bowl to a bread basket"[27] by the Green Revolution and food surpluses put an end to India's living a "ship-to-mouth" existence. This common belief is based on the impression that food grain imports after the Green Revolution substantially declined. In fact, however, food imports have continued to be significant even after the Green Revolution, as illustrated in table 2.5.

A second reason for the Green Revolution being seen as a miracle lies in an ahistorical view of grain trade. The flow of grain from North to South is of recent origin, before which, grain traveled from the South to the North. India was a major supplier of wheat to Europe until the war years. As Dan Morgan reports, "In 1873, with the opening of the Suez Canal, the first wheat arrived from India, after a push by British entrepreneurs to obtain a cheap, secure source of wheat under British control. The British envisaged India as a potentially secure source of wheat for the Empire.

Industrial tycoons pushed rail roads and canals into the Indus and Ganges river basins, where farmers had been growing wheat for centuries."[28]

According to George Blyn, in the quarter century before World War I, rising per capita output and consumption pervaded all major regions. "Most foodgrain crops also expanded at substantial rates, and though much rice and wheat were exported, domestic availability grew at about the same rate as output. . . . This early period gives evidence that per capita consumption of agricultural commodities increased over a substantial period of years."[29]

In times of crisis and scarcity, the colonial government of course put its revenue needs above those of the survival of the people. On November 3, 1772, a year after the great famine in Bengal that killed about 10 million people, Warren Hastings wrote to the Court of Directors of the East India Company: "Notwithstanding the loss of at least one third of the inhabitants of the province, and the consequent decrease of the cultivation the net collection of the year 1771 exceeded even those of 1768. . . . It was naturally to be expected that the diminution of the revenue should have kept an equal pace with the other consequences of so great a calamity. That it did not was owing to its being violently kept up to its former standard."[30]

Injustice has been at the root of the worst forms of scarcity throughout human history, and injustice and inequality have also been at the root of societal violence. By separating issues of agricultural production from issues of justice, the Green Revolution strategy attempted to defuse political turmoil. But bypassing the goals of equality and sustainability led to the creation of new inequalities and new scarcities. The Green Revolution strategy for peace had boomeranged. In creating new polarization, it created new potential for conflict. As Binswager and Futten noted:

> It does seem clear . . . that the contribution of the new seed fertilizer technology to food grain production has weakened the potential for revolutionary change in political and economic institutions in rural areas in many countries in Asia and in other parts of the developing world. In spite of widening income differentials, the gains in productivity growth, in those areas where the new seed-fertilizer technology has been effective, have been sufficiently diffused to preserve the vested interests of most

classes in an evolutionary rather than a revolutionary pattern of rural development.

By the mid-1970s, however, the productivity gains that had been achieved during the previous decade were coming more slowly and with greater difficulty in many areas. Perhaps revolutionary changes in rural institutions that the radical critics of the Green Revolution for the past ten years have been predicting will occur as a result of increasing immiserization in the rural areas of many developing countries during the coming decade.[31]

Notes

1. Jack Doyle, *Altered Harvest* (New York: Viking, 1985), 256.

2. S. Harding, *The Science Question in Feminism* (Ithaca: Cornell University Press, 1986), 30.

3. Vandana Shiva, "Reductionist Science as Epistemic Violence," in *Science, Hegemony and Violence,* ed. A. Nandy (Delhi: Oxford University Press, 1988).

4. Albert Howard, *An Agricultural Testament* (London: Oxford University Press, 1940).

5. John Augustus Voelcker, *Report on the Improvement of Indian Agriculture* (London: Eyre and Spottiswoode, 1893), 11.

6. M. K. Gandhi, *Food Shortage and Agriculture* (Ahmedabad: Najivan, 1949), 47.

7. K. M. Munshi, *Towards Land Transformation* (Ministry of Food and Agriculture, n.d.), 145.

8. C. Subramaniam, *The New Strategy in Agriculture* (New Delhi: Vikas, 1979).

9. Jaganath Pathy, "Green Revolution in India" (paper presented at seminar The Crisis in Agriculture, APPEN/TWN, Penang, January 1990).

10. E. Taboada, quoted by Gustavo Esteva, "Beyond the Knowledge/Power Syndrome: The Case of the Green Revolution" (paper presented at UNU/WIDER seminar, Karachi, January 1989), 19.

11. E. Hyams, *Soil and Civilisation* (London: Thames and Hudson, 1952).

12. A. S. Johnson, "The Foundations Involvement in Intensive Agricultural Development in India," in *Cropping Patterns in India* (New Delhi: ICAR, 1978).

13. Doyle, *Altered Harvest,* 256.

14. Claude Alvares, "The Great Gene Robbery," *Illustrated Weekly of India,* March 23, 1986.

15. B. Onate, "Why the Green Revolution Has Failed the Small Farmers" (paper presented at CAP seminar Problems and Prospects of Rural Malaysia, Penang, November 1985).

16. Frances Moore Lappe and Joseph Collins, *Food First* (London: Abacus, 1982), 114.

17. Angus Wright, "Innocents Abroad: American Agricultural Research in Mexico," in *Meeting the Expectations of the Land,* ed. Wes Jackson et al. (San Francisco: North Point, 1984).

18. R. P. Dutt, quoted in J. Bajaj, "Green Revolution: A Historical Perspective" (paper presented at CAP/TWN seminar The Crisis in Modern Science, Penang, November 1986), 4.

19. Ibid.

20. Esteva, "Beyond the Knowledge/Power Syndrome," 19.

21. Lappe and Collins, *Food First,* 114.

22. Robert Anderson and Baker Morrison, *Science, Politics and the Agricultural Revolution in Asia* (Boulder: Westview, 1982), 7.

23. Ibid., 5.

24. Ibid., 3.

25. Harry Cleaver, "Technology as Political Weaponry," in ibid., 269.

26. David Hopper, quoted in Andrew Pearse, *Seeds of Plenty, Seeds of Want* (Oxford: Oxford University Press, 1980), 79.

27. M. S. Swaminathan, *Science and the Conquest of Hunger* (Delhi: Concept, 1983), 409.

28. Dan Morgan, *Merchants of Grain* (New York: Viking, 1979), 36.

29. George Blyn, "India's Crop Output Trends, Past and Present," in *Agricultural Development of India: Policy and Problems,* ed. C. M. Shah (Delhi: Orient Longman, 1979), 583.

30. Quoted in Bajaj, "Green Revolution," 5.

31. Quoted in Edmund Oasa, "The Political Economy of International Agricultural Research: A Review of the CGIAR's Response to Criticisms of the Green Revolution," in *The Green Revolution Revisited,* ed. B. Gleaser (Boston: Allen and Unwin, 1956), 25.

3

The Hijacking of
the Global Food Supply

Food is our most basic need, the very stuff of life. According to an ancient Indian Upanishad, "All that is born is born of *anna* [food]. Whatever exists on earth is born of *anna*, lives on *anna*, and in the end merges into *anna*. *Anna* indeed is the first born amongst all beings."[1]

More than 3.5 million people starved to death in the Bengal famine of 1943. Twenty million were directly affected. Food grains were appropriated forcefully from the peasants under a colonial system of rent collection. Export of food grains continued in spite of the fact that people were going hungry. As the Bengali writer Kali Charan Ghosh reports, eighty thousand tons of food grain were exported from Bengal in 1943, just before the famine. At the time, India was being used as a supply base for the British military. "Huge exports were allowed to feed the people of other lands, while the shadow of famine was hourly lengthening on the Indian horizon."[2]

More than one-fifth of India's national output was appropriated for war supplies. The starving Bengal peasants gave up over two-thirds of the food they produced, leading their debt to double. This, coupled with speculation, hoarding, and profiteering by traders, led to skyrocketing prices. The poor of Bengal paid for the empire's war through hunger and starvation—and the "funeral march of the Bengal peasants, fishermen, and Artisans."[3] Dispossessed peasants moved to Calcutta. Thousands of female destitutes were turned into prostitutes. Parents started to sell their children. "In the villages jackals and dogs engaged in a tug-of-war for the bodies of the half-dead."[4]

As the crisis began, thousands of women organized in Bengal in defense of their food rights. "Open more ration shops" and "Bring down the price of food" were the calls of women's groups throughout Bengal.[5]

41

After the famine, the peasants also started to organize around the central demand of keeping a two-thirds, or *tebhaga,* share of the crops. At its peak, the Tebhaga Movement, as it was called, covered nineteen districts and involved 6 million people. Peasants refused to let their harvest be stolen by the landlords and the revenue collectors of the British Empire. Everywhere peasants declared, "Jan debo tabu dhan debo ne"—"We will give up our lives, but we will not give up our rice." In the village of Thumniya, the police arrested some peasants who resisted the theft of their harvest. They were charged with "stealing paddy."[6]

A half century after the Bengal famine, a new and clever system has been put in place, which is once again making the theft of the harvest a right and the keeping of the harvest a crime. Hidden behind complex free trade treaties are innovative ways to steal nature's harvest, the harvest of the seed, and the harvest of nutrition.

The Corporate Hijacking of Food and Agriculture

I focus on India to tell the story of how corporate control of food and globalization of agriculture are robbing millions of their livelihoods and their right to food both because I am an Indian and because Indian agriculture is being especially targeted by global corporations. Since 75 percent of the Indian population derives its livelihood from agriculture, and every fourth farmer in the world is an Indian, the impact of globalization on Indian agriculture is of global significance.

However, this phenomenon of the stolen harvest is not unique to India. It is being experienced in every society, as small farms and small farmers are pushed to extinction, as monocultures replace biodiverse crops, as farming is transformed from the production of nourishing and diverse foods into the creation of markets for genetically engineered seeds, herbicides, and pesticides. As farmers are transformed from producers into consumers of corporate-patented agricultural products, as markets are destroyed locally and nationally but expanded globally, the myth of "free trade" and the global economy becomes a means for the rich to rob the poor of their right to food and even their right to life. For the vast majority of the world's people—70 percent—earn their livelihoods by producing food. The majority of these farmers are women. In contrast, in the industrialized countries, only 2 percent of the population are farmers.

Food Security Is in the Seed

For centuries third-world farmers have evolved crops and given us the diversity of plants that provide us nutrition. Indian farmers evolved two hundred thousand varieties of rice through their innovation and breeding. They bred rice varieties such as basmati. They bred red rice and brown rice and black rice. They bred rice that grew eighteen feet tall in the Gangetic floodwaters, and saline-resistant rice that could be grown in the coastal waters. And this innovation by farmers has not stopped. Farmers involved in our movement, Navdanya, dedicated to conserving native seed diversity, are still breeding new varieties.

The seed, for the farmer, is not merely the source of future plants and food; it is the storage place of culture and history. Seed is the first link in the food chain. Seed is the ultimate symbol of food security.

Free exchange of seed among farmers has been the basis of maintaining biodiversity as well as food security. This exchange is based on cooperation and reciprocity. A farmer who wants to exchange seed generally gives an equal quantity of seed from his field in return for the seed he gets. Free exchange among farmers goes beyond mere exchange of seeds; it involves exchanges of ideas and knowledge, of culture and heritage. It is an accumulation of tradition, of knowledge of how to work the seed. Farmers learn about the plants they want to grow in the future by watching them grow in other farmers' fields.

Paddy, or rice, has religious significance in most parts of the country and is an essential component of most religious festivals. The Akti festival in Chattisgarh, where a diversity of *indica* rices is grown, reinforces the many principles of biodiversity conservation. In southern India, rice grain is considered auspicious, or *akshanta*. It is mixed with *kumkum* and turmeric and given as a blessing. The priest is given rice, often along with coconut, as an indication of religious regard. Other agricultural varieties whose seeds, leaves, or flowers form an essential component of religious ceremonies include coconut, betel, areca nut, wheat, finger and little millets, horse gram, black gram, chickpea, pigeon pea, sesame, sugarcane, jackfruit seed, cardamom, ginger, bananas, and gooseberry.

New seeds are first worshipped, and only then are they planted. New crops are worshipped before being consumed. Festivals held before sowing seeds as well as harvest festivals, celebrated in the fields, symbolize people's intimacy with nature.[7] For the farmer, the field is the mother; worshipping the field is a sign of gratitude toward the earth, which, as mother, feeds the millions of life-forms that are her children.

But new intellectual property rights regimes, which are being universalized through the Trade Related Intellectual Property Rights Agreement of the World Trade Organization (WTO), allow corporations to usurp the knowledge of the seed and monopolize it by claiming it as their private property. Over time, this results in corporate monopolies over the seed itself.

Corporations like RiceTec of the United States are claiming patents on basmati rice. Soybean, which evolved in East Asia, has been patented by Calgene, which is now owned by Monsanto. Calgene also owns patents on mustard, a crop of Indian origin. Centuries of collective innovation by farmers and peasants are being hijacked as corporations claim intellectual property rights on these and other seeds and plants.[8]

"Free Trade" or "Forced Trade"?

Today, ten corporations control 32 percent of the commercial seed market, valued at $23 billion, and 100 percent of the market for genetically engineered, or transgenic, seeds.[9] These corporations also control the global agrochemical and pesticide market. Just five corporations control the global trade in grain. In late 1998, Cargill, the largest of these five companies, bought Continental, the second largest, making it the single biggest factor in the grain trade. Monoliths such as Cargill and Monsanto were both actively involved in shaping international trade agreements, in particular the Uruguay Round of the General Agreement on Trade and Tariffs, which led to the establishment of the WTO.

This monopolistic control over agricultural production, along with structural adjustment policies that brutally favor exports, results in floods of exports of foods from the United States and Europe to the third world. As a result of the North American Free Trade Agreement (NAFTA), the proportion of Mexico's food supply that is imported has increased from 20 percent in 1992 to 43 percent in 1996. After eighteen months of NAFTA, 2.2. million Mexicans have lost their jobs, and 40 million have fallen into extreme poverty. One out of two peasants is not getting enough to eat. As Victor Suares has stated, "Eating more cheaply on imports is not eating at all for the poor in Mexico."[10]

In the Philippines, sugar imports have destroyed the economy. In Kerala, India, the prosperous rubber plantations were rendered unviable due to rubber imports. The local $350 million rubber economy was wiped out, with a multiplier effect of $3.5 billion on the economy of Kerala. In Kenya, maize imports brought prices crashing for local farmers, who could not even recover their costs of production.

Trade liberalization of agriculture was introduced in India in 1991 as part of a World Bank/International Monetary Fund (IMF) structural adjustment package. While the hectares of land under cotton cultivation had been decreasing in the 1970s and 1980s, in the first six years of World Bank/IMF–mandated reforms, the land under cotton cultivation increased by 1.7 million hectares. Cotton started to displace food crops. Aggressive corporate advertising campaigns, including promotional films shown in villages on "video vans," were launched to sell new hybrid seeds to farmers. Even gods, goddesses, and saints were not spared: in Punjab, Monsanto sells its products using the image of Guru Nanak, the founder of the Sikh religion. Corporate hybrid seeds began to replace local farmers' varieties.

The new hybrid seeds, being vulnerable to pests, required more pesticides. Extremely poor farmers bought both seeds and chemicals on credit from the same company. When the crops failed due to heavy pest incidence or large-scale seed failure, many peasants committed suicide by consuming the same pesticides that had gotten them into debt in the first place. In the district of Warangal, nearly four hundred cotton farmers committed suicide due to crop failure in 1997, and dozens more committed suicide in 1998.

Under this pressure to cultivate cash crops, many states in India have allowed private corporations to acquire hundreds of acres of land. The state of Maharashtra has exempted horticulture projects from its land-ceiling legislation. Madhya Pradesh is offering land to private industry on long-term leases, which, according to industry, should last for at least forty years. In Andhra Pradesh and Tamil Nadu, private corporations are today allowed to acquire over three hundred acres of land for raising shrimp for exports. A large percentage of agricultural production on these lands will go toward supplying the burgeoning food-processing industry, in which mainly transnational corporations are involved. Meanwhile, the United States has taken India to the WTO dispute panel to contest its restrictions on food imports.

In certain instances, markets are captured by other means. In August 1998, the mustard oil supply in Delhi was mysteriously adulterated. The adulteration was restricted to Delhi but not to any specific brand, indicating that it was not the work of a particular trader or business house. More than fifty people died. The government banned all local processing of oil and announced free imports of soybean oil. Millions of people extracting oil on tiny, ecological, cold-press mills lost their livelihoods. Prices of indigenous oilseed collapsed to less than one-third their previous levels. In Sira, in the state of Karnataka, police officers shot farmers protesting the fall in prices of oilseeds.

Imported soybeans' takeover of the Indian market is a clear example of the imperialism on which globalization is built. One crop exported from a single country by one or two corporations replaced hundreds of foods and food producers, destroying biological and cultural diversity, and economic and political democracy. Small mills are now unable to serve small farmers and poor consumers with low-cost, healthy, and culturally appropriate edible oils. Farmers are robbed of their freedom to choose what they grow, and consumers are robbed of their freedom to choose what they eat.

Creating Hunger with Monocultures

Global chemical corporations, recently reshaped into "life sciences" corporations, declare that without them and their patented products, the world cannot be fed. As Monsanto advertised in its $1.6 million European advertising campaign:

> Worrying about starving future generations won't feed them. Food biotechnology will. The world's population is growing rapidly, adding the equivalent of a China to the globe every ten years. To feed these billion more mouths, we can try extending

our farming land or squeezing greater harvests out of existing cultivation. With the planet set to double in numbers around 2030, this heavy dependency on land can only become heavier. Soil erosion and mineral depletion will exhaust the ground. Lands such as rainforests will be forced into cultivation. Fertilizer, insecticide, and herbicide use will increase globally. At Monsanto, we now believe food biotechnology is a better way forward.[11]

But food is necessary for all living species. That is why the *Taittreya Upanishad* calls on humans to feed all beings in their zone of influence. Industrial agriculture has not produced more food. It has destroyed diverse sources of food, and it has stolen food from other species to bring larger quantities of specific commodities to the market, using huge quantities of fossil fuels and water and toxic chemicals in the process.

It is often said that the so-called miracle varieties of the Green Revolution in modern industrial agriculture prevented famine because they had higher yields. However, these higher yields disappear in the context of total yields of crops on farms. Green Revolution varieties produced more grain by diverting production away from straw. This "partitioning" was achieved through dwarfing the plants, which also enabled them to withstand high doses of chemical fertilizer.

However, less straw means less fodder for cattle and less organic matter for the soil to feed the millions of soil organisms that make and rejuvenate soil. The higher yields of wheat or maize were thus achieved by stealing food from farm animals and soil organisms. Since cattle and earthworms are our partners in food production, stealing food from them makes it impossible to maintain food production over time, meaning that the partial yield increases were not sustainable.

The increase in yields of wheat and maize under industrial agriculture were also achieved at the cost of yields of other foods a small farm provides. Beans, legumes, fruits, and vegetables all disappeared both from farms and from the calculus of yields. More grain from two or three commodities arrived on national and international markets, but less food was eaten by farm families in the third world.

The gain in "yields" of industrially produced crops is thus based on a theft of food from other species and the rural poor in the third world. That is why, as more grain is produced and traded globally, more people go hungry in the third world. Global markets have more commodities for trading because food has been robbed from nature and the poor.

Productivity in traditional farming practices has always been high if it is remembered that very few external inputs are required. While the Green Revolution has been promoted as having increased productivity in the absolute sense, when resource use is taken into account, it has been found to be counterproductive and inefficient.

Perhaps one of the most fallacious myths propagated by Green Revolution advocates is the assertion that high-yielding varieties have reduced the acreage under cultivation, therefore preserving millions of hectares of biodiversity. But in India, instead of more land being released for conservation, industrial breeding actually increases pressure on the land, since each acre of a monoculture provides a single output, and the displaced outputs have to be grown on additional, or "shadow" acres.[12]

A study comparing traditional polycultures with industrial monocultures shows that a polyculture system can produce 100 units of food from 5 units of inputs, whereas an industrial system requires 300 units of input to produce the same 100 units. The 295 units of wasted inputs could have provided 5,900 units of additional food. Thus the industrial system leads to a decline of 5,900 units of food. This is a recipe for starving people, not for feeding them.[13] Wasting resources creates hunger. By wasting resources through one-dimensional monocultures maintained with intensive external inputs, the new biotechnologies create food insecurity and starvation.

The Insecurity of Imports

As cash crops such as cotton increase, staple food production goes down, leading to rising prices of staples and declining consumption by the poor. The hungry starve as scarce land and water are diverted to provide luxuries for rich consumers in Northern countries. Flowers, fruits, shrimp, and meat are among the export commodities being promoted in all third-world countries.

When trade liberalization policies were introduced in 1991 in

India, the agriculture secretary stated that "food security is not food in the *go-downs* but dollars in the pocket." It is repeatedly argued that food security does not depend on food "self-sufficiency" (food grown locally for local consumption) but on food "self-reliance" (buying your food from international markets). According to the received ideology of free trade, the earnings from exports of farmed shrimp, flowers, and meat will finance imports of food. Hence any shortfall created by the diversion of productive capacity from growing food for domestic consumption to growing luxury items for consumption by rich Northern consumers would be more than made up.

However, it is neither efficient nor sustainable to grow shrimp, flowers, and meat for export in countries such as India. In the case of flower exports, India spent Rs. 1.4 billion as foreign exchange for promoting floriculture exports and earned a mere Rs. 320 million.[14] In other words, India can buy only one-fourth of the food it could have grown with export earnings from floriculture.[15] Our food security has therefore declined by 75 percent, and our foreign exchange drain increased by more than Rs. 1 billion.

In the case of meat exports, for every $1 earned, India is destroying $15 worth of ecological functions performed by farm animals for sustainable agriculture. Before the Green Revolution, the by-products of India's culturally sophisticated and ecologically sound livestock economy, such as the hides of cattle, were exported rather than the ecological capital, that is, the cattle themselves. Today, the domination of the export logic in agriculture is leading to the export of our ecological capital, which we have conserved over centuries. Giant slaughterhouses and factory farming are replacing India's traditional livestock economy. When cows are slaughtered and their meat is exported, with it are exported the renewable energy and fertilizer that cattle provide to the small farms of small peasants. These multiple functions of cattle in farming systems have been protected in India through the metaphor of the sacred cow. Government agencies cleverly disguise the slaughter of cows, which would outrage many Indians, by calling it "buffalo meat."

In the case of shrimp exports, for every acre of an industrial shrimp farm, two hundred acres of productive ecosystems are destroyed. For every $1 earned as foreign exchange from exports, $6 to $10 worth of destruction takes place in the local economy. The harvest of shrimp from aquaculture farms is a harvest stolen from fishing and farming communities in the coastal regions of the third world. The profits from exports of shrimp to U.S., Japanese, and European markets show up in national and global economic growth figures. However, the destruction of local food consumption, groundwater resources, fisheries, agriculture, and livelihoods associated with traditional occupations in each of these sectors does not alter the global economic value of shrimp exports; such destruction is experienced only locally.

In India, intensive shrimp cultivation has turned fertile coastal tracts into graveyards, destroying both fisheries and agriculture. In Tamil Nadu and Andhra Pradesh, women from fishing and farming communities are resisting shrimp cultivation through satyagraha. Shrimp cultivation destroys fifteen jobs for each job it creates. It destroys $5 of ecological and economic capital for every $1 earned through exports. Even these profits flow for only three to five years, after which the industry must move on to new sites. Intensive shrimp farming is a nonsustainable activity, described by United Nations agencies as a "rape and run" industry.

Since the World Bank is advising all countries to shift from "food-first" to "export-first" policies, these countries all compete with each other, and the prices of these luxury commodities collapse. Trade liberalization and economic reform also include devaluation of currencies. Thus exports earn less, and imports cost more. Since the third world is being told to stop growing food and instead to buy food in international markets by exporting cash crops, the process of globalization leads to a situation in which agricultural societies of the South become increasingly dependent on food imports but do not have the foreign exchange to pay for imported food. Indonesia and Russia provide examples of countries that have moved rapidly from food sufficiency to hunger because of the creation of dependency on imports and the devaluation of their currencies.

Stealing Nature's Harvest

Global corporations are not just stealing the harvest of farmers. They are stealing nature's harvest through genetic engineering and patents on life-forms. Genetically engineered crops manufactured by corporations pose serious ecological risks. Crops such as Monsanto's Roundup Ready soybeans, designed to be resistant to herbicides, lead to the destruction of biodiversity and the increased use of agrochemicals. They can also create highly invasive "superweeds" by transferring the genes for herbicide resistance to weeds. Crops designed to be pesticide factories, genetically engineered to produce toxins and venom with genes from bacteria, scorpions, snakes, and wasps, can threaten nonpest species and can contribute to the emergence of resistance in pests and hence the creation of "superpests." In every application of genetic engineering, food is being stolen from other species for the maximization of corporate profits.

To secure patents on life-forms and living resources, corporations must claim seeds and plants to be their "inventions" and hence their property. Thus corporations like Cargill and Monsanto see nature's web of life and cycles of renewal as "theft" of their property. During the debate about the entry of Cargill into India in 1992, the Cargill chief executive stated, "We bring Indian farmers smart technologies, which prevent bees from usurping the pollen."[16] During the United Nations Biosafety Negotiations, Monsanto circulated literature claiming that "weeds steal the sunshine."[17] A worldview that defines pollination as "theft by bees" and claims that diverse plants "steal" sunshine is one aimed at stealing nature's harvest by replacing open, pollinated varieties with hybrids and sterile seeds, and destroying biodiverse flora with herbicides such as Monsanto's Roundup.

This is a worldview based on scarcity. A worldview of abundance is the worldview of women in India who leave food for ants on their doorstep, even as they create the most beautiful art in *kolams, mandalas,* and *rangoli* with rice flour. Abundance is the worldview of peasant women who weave beautiful designs of paddy to hang up for birds when the birds do not find grain in the fields. This view of abundance recognizes that, in giving food to other beings and species, we maintain conditions for our own food security. It is the recognition in the *Isho Upanishad* that the universe is the creation of the Supreme Power meant for the benefits of (all) creation. Each individual life-form must learn to enjoy its benefits by farming a part of the system in close relation with other species. Let not any one species encroach upon others' rights.[18] The *Isho Upanishad* also says, "A selfish man over-utilizing the resources of nature to satisfy his own ever-increasing needs is nothing but a thief, because using resources beyond one's needs would result in the utilization of resources over which others have a right."[19]

In the ecological worldview, when we consume more than we need or exploit nature on principles of greed, we are engaging in theft. In the antilife view of agribusiness corporations, nature renewing and maintaining itself is a thief. Such a worldview replaces abundance with scarcity, fertility with sterility. It makes theft from nature a market imperative, and hides it in the calculus of efficiency and productivity.

Food Democracy

What we are seeing is the emergence of food totalitarianism, in which a handful of corporations controls the entire food chain and destroys alternatives so that people do not have access to diverse, safe foods produced ecologically. Local markets are being deliberately destroyed to establish monopolies over seed and food systems. The destruction of the edible oil market in India and the many ways through which farmers are prevented from having their own seed supply are small instances of an overall trend in which trade rules, property rights, and new technologies are used to destroy people-friendly and environment-friendly alternatives and to impose antipeople, antinature food systems globally.

The notion of rights has been turned on its head under globalization and free trade. The right to produce for oneself or consume according to cultural priorities and safety concerns has been rendered illegal according to the new trade rules. The right of corporations to force-feed citizens of the world with culturally inappropriate and hazardous foods has been made absolute. The right to food, the right to safety, the right to culture are all being treated as trade barriers that need to be dismantled.

This food totalitarianism can be stopped only through major citizen mobilization for democratization of the food system. This mobilization is starting to gain momentum in Europe, Japan, India, Brazil, and other parts of the world.

We have to reclaim our right to save seed and to biodiversity. We have to reclaim our right to nutrition and food safety. We have to reclaim our right to protect the earth and its diverse species. We have to stop this corporate theft from the poor and from nature. Food democracy is the new agenda for democracy and human rights. It is the new agenda for ecological sustainability and social justice.

Notes

1. *Taittreya Upanishad* (Gorakhpur: Gita), 124.

2. Kali Charan Ghosh, *Famines in Bengal, 1770–1943* (Calcutta: Indian Associated Publishing, 1944).

3. Bondhayan Chattopadhyay, "Notes towards an Understanding of the Bengal Famine of 1943," *Transaction*, June 1981.

4. MARS (Mahila Atma Raksha Samiti, or Women's Self-Defense League), *Political Report Prepared for Second Annual Conference* (New Delhi: Research Foundation for Science, Technology, and Ecology [RFSTE], 1944).

5. Peter Custers, *Women in the Tebhaga Uprising* (Calcutta: Naya Prokash, 1987), 52.

6. Ibid., 78.

7. Festivals like Uganda, Ramanavami, Akshay Trateeya, Ekadashi Aluyana Amavase, Naga Panchami, Noolu Hunime, Ganesh Chaturthi, Rishi Panchami, Navartri, Deepavali, Rathasaptami, Tulsi Vivaha Campasrusti, and Bhoomi Puja all include religious ceremonies around the seed.

8. Vandana Shiva, Vanaja Ramprasad, Pandurang Hegde, Omkar Krishnan, and Radha Holla-Bhar, *The Seed Keepers* (New Delhi: Navdanya, 1995).

9. These companies are DuPont/Pioneer (United States), Monsanto (United States), Novartis (Switzerland), Groupe Limagrain (France), Advanta (United Kingdom and Netherlands), Guipo Pulsar/Semins/ELM (Mexico), Sakata (Japan), KWS HG (Germany), and Taki (Japan).

10. Victor Suares, paper presented at International Conference on Globalization, Food Security, and Sustainable Agriculture, July 30–31, 1996.

11. *Monsanto: Peddling "Life Sciences" or "Death Sciences"?* (New Delhi: RFSTE, 1998).

12. ASSINSEL (International Association of Plant Breeders), *Feeding the 8 Billion and Preserving the Planet* (Nyon, Switzerland: ASSINSEL).

13. Francesca Bray, "Agriculture for Developing Nations," *Scientific American*, July 1994, 33–35.

14. *Business India*, March 1998.

15. T. N. Prakash and Tejaswini, "Floriculture and Food Security Issues: The Case of Rose Cultivation in Bangalore," in *Globalization and Food Security: Proceedings of Conference on Globalization and Agriculture*, ed. Vandana Shiva (New Delhi, 1996).

16. Interview with John Hamilton, *Sunday Observer*, May 9, 1993.

17. Hendrik Verfaillie, speech delivered at the Forum on Nature and Human Society, National Academy of Sciences, Washington, DC, October 30, 1997.

18. Vandana Shiva, *Globalization, Gandhi, and Swadeshi: What Is Economic Freedom? Whose Economic Freedom?* (New Delhi: RFSTE, 1998).

19. Ibid.

4

Hunger by Design

Why is every fourth Indian hungry? Why is every third woman in India anemic and malnourished? Why is every second child underweight, stunted, and wasted? Why has the hunger and malnutrition crisis deepened even as India has seen 9 percent growth? Why is "shining India" a starving India?

In my view, hunger is a structural part of the design of the Green Revolution, a design for scarcity. There is now much talk of a second "Green Revolution" in India and a "Green Revolution" in Africa. The second Green Revolution is based on genetic engineering, which is being introduced into agriculture largely to allow corporations to claim intellectual property rights and patents on seeds. The floodgate of patenting seeds was opened through the Trade Related Intellectual Property Rights Agreement (TRIPS) of WTO, written by corporations like Monsanto.

The Agreement on Agriculture (AoA) of the WTO was drafted by the multinational corporation (MNC) Cargill, designed to allow it and other agribusiness corporations to have access to world markets by forcing countries to remove import restrictions (quantitative restrictions) and using $400 billion to subsidize and dump artificially cheap food commodities on countries of the South. The dumping of soy and the destruction of India's domestic edible oil production and distribution is an example of the global reach of MNCs; Indian farmers are losing $25 billion every year to falling prices but food prices continue to rise, creating a double burden of hunger for rural communities, which is why half the hungry people in India and the world are farmers.

Globalized forced trade in food, falsely called free trade, has aggravated the hunger crisis by undermining food sovereignty and food democracy. With the deadlock in the Doha round of WTO, forced trade is being driven by bilateral agreements such as the U.S.-India Knowledge Initiative in Agriculture, on whose board sit corporations like Monsanto, Cargill/ADM, and Walmart. Sadly, India is trying to use the food crisis that trade liberalization policies have created to hand over seed supply to Monsanto, food supply to Cargill and other corporations, and retail to Walmart, in line with the U.S.-India Agreement on Agriculture signed with President Bush in 2005. Speaking at a conference on the crisis and food inflation on February 4, 2011, Prime Minister Manmohan Singh said India needs to "shore up farm supply chains by bringing in organised retail players" (read Walmart).[1] Recent research shows that globalized retail is destroying farmers' livelihoods, destroying livelihoods in small retail, and leading to wastages of up to 50 percent of food. This, too, is hunger by design.

Both the U.S. and Indian governments are supporting U.S. agribusiness corporations to expand markets and profits; the common citizen is politically orphaned in a world shaped by corporate rule—farmers' rights and people's right to food are extinguished. When the Supreme Court of India told the government to distribute the food grain that was rotting in its godowns, the prime minister said he could not do so because it would distort "the market." When the National Advisory Council (NAQ) headed by Sonia Gandhi drafted a Food Security Act, the prime minister's appointed Rangarajan Committee said it would distort "the market." In other words, corporate rights to profit through the creation of hunger must be protected even as people die.

And even as corporate greed has led to the food crisis, the corporate takeover of seed, food, and land is being offered up as a solution to it. The government has already allowed 2 million hectares of fertile farmland to go out of food production; new farmland is being given over to agribusinesses. Planning Commission vice chairman Montek Singh Ahluwalia, during a visit to Muscat, invited Gulf countries to farm in India and export food to their countries.[2] A Bahrain firm, the Nader and Ebrahim Group (NEG), recently tied up with the Pune-based Sanghar Group to grow bananas on four hundred acres; so far 2.6 million kilos of bananas have been exported. Indian laws do not allow foreigners to buy land; the Planning Commission is encouraging foreign corporations to subvert India's land sovereignty by asking them to enter into partnerships with Indian companies and by encouraging contract farming.

Diverting land from food for local communities to cash crops for the rich in United States, Europe, and the Gulf can only aggravate the food crisis. Biodiverse organic farming, if adopted nationally, can provide enough calories for 2.4 billion people, enough protein for 2.5 billion, enough carotene for 1.5 billion, and enough folic acid for 1.7 billion pregnant women. There is no place for hunger in a sustainable, just, and democratic society. We must end it by building food democracy, by reclaiming our seed sovereignty, food sovereignty, and land sovereignty.

India Shining or India Starving?

Per capita consumption today has dropped from 177 kilograms a day to 152 kilograms a day as a result of the food chain being broken; tinkering with fragments of the broken chain will not fix it. The first link in the food chain begins with the natural capital of soil, water, and seed; the second link is the labor of small, marginal farmers and landless peasants, most of whom are women; the final link is eating. The first link has been broken by ecological degradation—soil erosion, biodiversity erosion, water depletion, undernourished food production contribute to food insecurity. When peasants lose access to land, seed, and water, they lose access to food; an increase in hunger is a direct consequence. The second link that has been broken is the capacity of the farmer to produce food. Rising costs of production and falling farm prices create debt and this creates food insecurity. The deliberate destruction of food procurement by dismantling the public distribution system, by using godowns to store liquor instead of food, and by not guaranteeing a fair price to farmers is a signal that the government wants a food system without small farmers. Farmers are the backbone of India's food security and food sovereignty, and there can be no food security in a deepening agrarian crisis. The third link in the food chain is people's entitlement and right to food. Rising food prices and decreasing production of pulses and nutritious millets have reduced the access of the poor to adequate food and nutrition.

While millions of our fellow citizens starve, the government fiddles with figures and addresses a fragment of the consequences of the crisis. Poverty is a consequence, not a cause. Fiddling with poverty figures—37 percent in the Tendulkar Committee Report, 50 percent in the Saxena Report, 77 percent in the Unorganised Sector Report—is a deliberate attempt to avoid addressing the root causes of hunger and poverty. In this context, the proposed National Food Security Act (NFSA) is a mere fig leaf. It is inadequate because it ignores the first two links in the food chain and reduces the scope of existing schemes for the poor and vulnerable. The NFSA offers only 25 kilograms of grain instead of the 35 kilograms per family, per month, fixed by the Supreme Court. The Indian Council of Medical Research has fixed the caloric norms at 2,400 kilocalories in rural areas and 2,100 kilocalories in urban areas; the Tendulkar Committee (which is now the Planning Commission's official basis) fixes average calorie consumption at 1,776 kilocalories in urban areas and 1,999 kilocalories in rural areas. Through juggling figures the hungry become well fed, the poor become nonpoor.

Food security demands a universal public distribution system (PDS) that serves both poor farmers and poor consumers by *ensuring fair prices throughout the food chain*. Instead, the government is committed to ever-narrowing "targets" because it is committed to handing over agriculture to global agribusiness, and handing over so-called food security schemes to companies like Walmart and Sodexo, Cargill, Unilever, and Nestlé, through introducing cash transfers. This is undermining food security by abandoning the farmer and the public distribution system.

As small farmers are displaced by agribusinesses, the destruction of natural capital will increase, weakening the first link in the food chain, and the agrarian crisis facing two-thirds of rural India will deepen. Breaking the link between farmers and eaters, between production and consumption through food stamps and food vouchers will completely break the food chain. The proposed solution is to reduce food subsidies; when PDS was replaced by targeted PDS under World Bank pressure, this was the argument used. However, the food subsidy bill increased from Rs. 2,500 crores to Rs. 50,000 crores—increased financially, but shrank socially from universal coverage to targeted PDS, thus effectively starving a quarter of our people. A further narrowing of the "target" will further increase the food subsidy because it will lead to an increase in the gap between the high cost of production and the cost of subsidized food as well as a growing gap between rising market prices for food and financing subsidized food. Privatizing our public food distribution system through cash transfers is a recipe for debt and hunger. Dismantling trade patterns that serve communities and replacing them with so-called free trade, which only increases market control and profits for agribusinesses, is akin to putting precious food in a global casino.

Why Are 1 Billion People Starving?

The year 2008 witnessed a global food crisis, with food prices rising to unprecedented levels and food riots taking place in forty countries; in 2010, the food crisis resurfaced. President Bush had an interesting analysis of the global rise in food prices in 2008. At an interactive session on the economy in Missouri, Bush argued that prosperity in countries like India had triggered an increased demand for better nutrition. "There are 350 million people in India who are classified as middle class. That's bigger than America. Their middle class is larger than our entire population. And when you start getting wealth, you start demanding better nutrition and better food, so demand is high and that causes prices to go up."

While this story might divert the U.S. political debate from the role of U.S. agribusiness in the current food crisis, both through speculation and through diversion of food to biofuels, and it might present economic globalization as having benefited Indians, the reality is that Indians are nutritionally worse off today than they were before globalization.

However, it was not just George Bush who blamed India for the price rise. Oxford economist Paul Collier, in his book *The Plundered Planet,* has stated, "The root cause of the sudden spike in prices was the spectacular economic growth of Asia. Asia is half the world. As Asian incomes rise, so too, does demand for food. Not only are Asians eating more, they are eating better."[3] It is being stated repeatedly that food prices rise due to "surging demand in emerging economies like China and India." This growth myth is false on many counts. First, while the Indian economy has indeed grown, the majority of Indians have grown poorer because they have lost their land and livelihoods. Most Indians are in fact eating less today than a decade ago; the per capita availability of food has declined from 177 kilograms per person, per year in 1991 to 152 kilograms per person, per year currently. The daily availability of food has declined from 485 to 419 grams per day; daily calorie intake has dropped from 2,220 calories per day to 2,150 calories per day. One million children die every year for lack of food, and even as India's growth soars, it has emerged as the capital of hunger. India ranks sixty-seventh among eighty-four countries in the global hunger index, below China and Pakistan, and is home to 42 percent of the world's under-weight children.

The poor are worse off because their food and livelihoods have been destroyed. The middle classes are worse off because they are eating worse, not better, as junk food and processed food is forced on India through globalization. India is passing through a dietary or nutritional transition, with malnutrition taking on a double face: on the one hand, malnutrition based on food deprivation, on the other, malnutrition linked to a junk-food diet and its equally debilitating health effects. The Indian middle class is eating less cereal today. In 1972–1973, urban Indians spent 23 percent on cereals; this is down to 10 percent because of U.S. pressure through the U.S.-India Agriculture Agreement to promote processed and packaged foods. Cereal consumption in the United States has grown by 12 percent, compared to 2 percent in India, largely as a result of diversion of food for biofuel. President Bush's biofuel policies and the deregulation of the financial economy, which has allowed speculation to enter the food economy, are the real reasons for food prices rising.

According to the U.N. special rapporteur on the right to food, Olivier de Schutter, 10 percent of the world's hungry are pastoralists, fisherfolk, or forest dwellers; 20 percent are the rural landless, such as farm laborers; and 50 percent are, in fact, smallholding farmers. If 80 percent of the world's hungry are producers of food, they are clearly not eating enough of what they produce. Hunger creation is built into the design of the unfree rules of "free trade," which turn farmers into seed slaves of the gene giants who patent and own seeds, and into indebted bonded labor of the grain giants who sell costly inputs and buy cheap commodities from farmers trapped in debt. Hunger creation is also built into the design of industrial processing, long-distance transport, and large-volume retail. Even while countries like India are told that our short distribution chains waste food, the reality is that the industrial globalized food system *wastes 50 percent of food produced.*

We have been told repeatedly that industrial agriculture is necessary to feed the world, that without the chemicals of the so-called Green Revolution and without genetically engineered organisms, the world will starve. We were told that free trade would make food cheap.

The food and hunger crisis is rooted in who owns natural capital—land, seeds and biodiversity, and water. It is determined by how we produce our food and how we distribute it. In the Indian context, agriculture, food, and nutrition are addressed independently of each other, even though the food that is grown determines its nutritional value, its distribution patterns and entitlements. If we grow biodiversity, including millets and pulses, we will have more nutrition per capita; if we grow monocultures with chemicals, we will have less nutrition per acre and per capita. If we grow food ecologically with internal inputs, more food will stay with the farming household and there will be less malnutrition in rural children. If we have local community-controlled food systems, we can escape the volatility of the global market and the monopolies of global corporations. If we have globalized trade in agriculture and speculative trade in food commodities, we will have debt and farmers' suicides, price rise and a food crisis, hunger and famine. We can create either hunger or food sovereignty by design; the former is shaped by a food dictatorship, the latter by food democracy.

Food in the Global Casino

Growing food, processing, transforming, and distributing it involves 70 percent of humanity; eating food involves all of us. Yet it is not culture or human rights that are shaping today's dominant food economy; it is speculation and profits. Putting food in the global financial casino is a design for hunger. After the U.S. subprime crisis and the Wall Street crash, investors rushed to commodity markets, especially oil and agricultural commodities. While real production did not increase between 2005 and 2007, commodity speculation in food increased 160 percent. Speculation pushed up prices and pushed an additional 100 million people to hunger. A 2008 advertisement for Deutsche Bank stated: "Do you enjoy rising prices? Everybody talks about commodities—with the Agriculture Euro Fund you can benefit from the increase in the value of the seven most important agricultural commodities."[4]

The financial deregulation that destabilized the world's financial system is now destabilizing the world's food system. Between 2003 and 2008, commodity index speculation increased by 1,900 percent, from $13 billion to $260 billion; 30 percent of these index funds were invested in food commodities. As the Agribusiness Accountability Initiative states, "We live in a brave new world of 24-hour electronic trading, triggered by algorithms of composite price indices, fits of investor 'lack of confidence' and of unregulated 'dark pools' of more than US $7 trillion in over-the-counter commodities derivatives trades."[5]

The world's commodity trading has no relationship to food, to its diversity, to its growers or eaters, to the seasons, to sowing or harvesting. Food diversity is reduced to eight commodities and bundled into "composite price indices," seasons are replaced by twenty-four-hour trading, food production driven by sunshine and photosynthesis is displaced by "dark pools of investment." The tragedy is that this unreal world is creating hunger for real people in the real world. In a cover story for *Harper's* magazine, Frederick Kaufman wrote about the food bubble: "How Wall Street starved millions and got away with it. The history of food took an ominous turn in 1991, at a time when no one was paying much attention. That was the year Goldman Sachs decided our daily bread might make an excellent investment." The entry of investors like Goldman Sachs, AIG Commodity Index, Bear Sterns, Oppenheimer Puneo, and Barclays allowed agribusiness to increase its profits. In the first quarter of 2008, Cargill attributed its 86 percent jump in profits to commodity trading, and Conagra sold its trading arm to a hedge fund for $2.8 billion.

Gambling on the price of wheat for profits took food away from 250 million people. As Austin Damani told Fred Kaufman, "We're trading wheat, but it's wheat we're never going to see, it's a cerebral experience." Food is an ecological, sensory, biological experience; with speculation it has been removed from its own reality. Grain markets have been transformed with futures trading by the grain giants in Chicago, Kansas City, and Minneapolis combined with speculation by investors. Kaufman says, "Imaginary wheat bought anywhere affects real wheat bought everywhere."[6] And if we do not decommodify food, more and more people will be denied food, as more and more money is poured into the global casino for profits.

The spike in world food prices began to reappear in 2011. According to the FAO, in January 2011, the food price index was up 3.4 percent from December 2010; the cereal price index was 3 percent above its December reading and at the highest level since July 2008, though still 11 percent below its peak in April 2008. The oils and fats index rose by 5.6 percent, approximating the June 2008 record level. The dairy price index shot up 6.2 percent and the sugar price index by 5.4 percent. Wheat prices were up by 25 percent compared to six months earlier. Prices of soybean and palm oil doubled over the second half of 2010.

In India, the price of onions jumped from Rs. 11 per kilogram in June 2010 to Rs. 75 per kilogram in January 2011. While the production of onions had gone up from 4.8 million tons in 2001–2002 to 12 million tons in 2009–2010, prices also went up, indicating that in a speculation-driven market there is no correlation between production and prices. The price difference between wholesale and retail prices was 135 percent. Tomato prices shot up by more than 100 percent between October and December 2010, from Rs. 15 per kilogram to Rs. 40–50 per kilogram. Prices of cabbage went up by 159 percent; garlic, 140 percent; potato, 86 percent; brinjal, 72 percent; and green peas, 66 percent between March and December 2010. While traders gained, farmers were losing. Farmers got only Rs. 8 per kilogram for tomatoes selling at Rs. 50. The price of staples has also been going up systematically. Between December 2006 and December 2010, rice went from Rs. 14.50 per kilogram to Rs. 24 per kilogram, sugar from Rs. 21 per kilogram to Rs. 34 per kilogram, arhar dal from Rs. 32 per kilogram to Rs. 65 per kilogram, moong dal from Rs. 46.50 per kilogram to Rs. 64 per kilogram.[7]

Synthetic Biology and Biodiversity Wars

Synthetic biology is the emerging technology for transforming biomass and biodiversity into commerce. Synthetic biology is an industry that creates "designer organisms to act as living factories." With synthetic biology, hopes are that by building biological systems from the ground up, they will function like computers or factories. The goal is to make biology easier to engineer by using biobricks.

Agriculture and food production were transformed through two earlier "Green Revolutions," the first based on introducing chemistry to agriculture. For this, plants were made into dwarf varieties with the application of chemical fertilizers to prevent them from lodging. The second "Green Revolution" is based on the application of genetic engineering; and the emerging third "Green Revolution" introduces synthetic biology.

All three are based on an inappropriate and outmoded mechanistic paradigm. Living systems are based on self-organization, diversity, and complexity; Green Revolutions reduce life to raw material, complex systems to "machines," diversity to monocultures. The Green Revolution defined plants as factories, running on inputs from other factories producing synthetic fertilizers. Biotechnology, the second Green Revolution, is based on the obsolete paradigm of genetic reductionism, of genes as "atoms" of plants, at a time when we know genes do not act in isolation and a single gene does not carry one trait but multiple traits like yield and resilience. Synthetic biology is even further removed from life.

To assess the promises and shortfalls of the emerging third "Green Revolution," we need to look at what the first and second Green Revolutions have delivered and the lessons that can be learned from them. The first false claim of both earlier revolutions is that they are miracles. As Angus Wright has observed with respect to the Green Revolution, "One way in which agricultural research went wrong was in saying, and allowing it to be said, that some miracle was being produced. . . . Historically, science and technology made their first advances by rejecting the idea of miracles in the natural world. Perhaps it would be best to return to that position."[8]

The second false claim is that of exaggerated benefits. Both earlier Green Revolutions have been declared miracle solutions to hunger. In the case of the first, it is said that HYVs (high-yielding varieties) saved millions from famine. This is not true. First, as Dr. Palmer concluded in the United Nations Research Institute for Social Development's fifteen-nation study of the impact of new seeds, the term "high-yielding varieties" is a misnomer because it implies that the new seeds are high yielding in and of themselves. The distinguishing feature of the new seeds, however, is that they are highly responsive to certain key inputs such as fertilizers and irrigation. Palmer therefore suggested the term "high responsive varieties" (HRV) be used instead.

Further, by transforming biodiverse systems into monocultures and by replacing tall straw, high biomass production varieties to dwarf varieties, the output of food, nutrition, and biomass actually went down. Biodiverse ecological systems produce more food and nutrition per unit acre, as Navdanya's study *Health per Acre* has shown. For example, in just one biodiverse system the following was observed. Overall, in energy terms, industrial agriculture is a negative energy system, using ten units of input to produce one unit of output. Industrial agriculture in the United States uses 380 times more energy per hectare to produce rice than a traditional farm in the Philippines; and energy use per kilo of rice is 80 times higher in the United States than in the Philippines. Energy use for maize production in the United States is 176 times more per hectare than a traditional farm in Mexico, and 33 times more per kilogram.

The first Green Revolution spread monocultures of rice, wheat, and corn. The second Green Revolution has spread monocultures of corn, soy, canola, and cotton. The third Green Revolution will spread monocultures for biofuels. Biomass advocates refer to "marginal," "unproductive," "idle," "degraded," and "abandoned lands and wastelands" as the target for biomass extraction. European researchers have said "a prerequisite for the bio-energy potential in all regions is that the present inefficient and low-intensive agricultural management systems are replaced by 2050 by the best practice agricultural management systems and technologies."

The industrialization of life is being sold as the new "bioeconomy," "clean tech," and the "green economy." While biodiversity economies are genuine bioeconomies, they are being eclipsed by the rise of bioeconomy as an industrial order based on biological materials and industrial processes. International organizations are already referring to it as follows: the OECD calls it bio-based economy; the European Union refers to it as knowledge-based bioeconomy; the World Economic Forum calls it a biorefinery industry; the Biotechnology Industry Organisation calls it biotechnology; while UNEP calls it the green economy, and the U.S. government's Biomass Research and Development Board calls it the bioeconomic revolution.

The ETC report *Biomassters* quotes Craig Venter, the founder of synthetic genomics, as saying, "Whoever produces abundant biofuels could end up making more than just big bucks—they will make history. The companies, the countries, that succeed in this will be the economic winners of the next age to the same extent that the oil-rich nations are today."[9]

The new bioeconomy imagined by Craig Venter is an extension of the oil economy. Instead of oil being mined from fossilized biological matter, it will now be squeezed out from living biological matter or biomass. Major players of the oil age are now engaged in a scramble for the biodiversity that performs ecological services in nature's economy and provides basic needs of food, fodder, fuel, fertilizer in people's sustenance economy. As the ETC report says, "With 24 per cent of the world's annual terrestrial biomass so far appropriated for human use, today's compounding crises are an opportunity to commodify and monopolise the remaining 76 per cent that Wall Street hasn't yet reached."

Industrial sectors with an interest in the biodiversity and biomass of the planet include the energy, chemical, plastics, food, textiles, pharmaceuticals, paper products, and building supplies industries. Along with carbon, this is a market of $17 trillion. Global corporations are joining in this earth-grab, including oil companies such as British Petroleum, Shell, Exxon Mobil; chemical and biotechnology companies like BASF and Dupont, Monsanto, Amyris, Synthetic Genomics, Syngenta; forestry and agribusiness companies like Cargill, Archer Daniel Midlands, Weyerhauser; food companies such as Procter & Gamble, Unilever, Coca-Cola; and financial giants Goldman Sachs, J. P. Morgan, Microsoft. As Craig Venter has said, "We have modest goals of replacing the whole petrochemical industry and becoming a major source of energy."

The 76 percent noncommercialized, noncommoditized biomass is the basis of biodiversity-based local economies. Biomass encompasses over 230 billion tons of "living stuff" that supports local living economies today and could support them in the future.

This annual bounty, known as earth's "primary production," is most abundant in the global South—in tropical oceans, forests, and fast-growing grasslands—sustaining the livelihoods, cultures and basic needs of most of the world's inhabitants. What is being sold as a benign and beneficial switch from black carbon to green carbon is, in fact, a red-hot resource-grab (from South to North) to capture a new source of wealth. If the grab succeeds, then plundering the biomassters of the South to cheaply run the industrial economies of the North will be an act of 21st century imperialism that deepens injustice and worsens poverty and hunger. Moreover, pillaging fragile ecosystems for their carbon and sugar stocks is a murderous move on an already overstressed planet.[10]

The biodiversity-based knowledge possessed by peasant communities to control pests and increase fertility is centuries old. The insecticidal use of nicotine, the alkaloid present in tobacco, dates back to the seventeenth century, long before it was isolated. The range of insects subject to control by nicotine is very wide, and the alkaloid has been used successfully by farmers against aphids, leaf rotters, moths, fruit tree borers, termites, cabbage butterfly larvae, and so on.

The neem (*azadirachta indica juss*), a large evergreen tree, is a native of India. The use of neem to ward off damage by pests has been known since antiquity, and farmers have always mixed neem leaves with grain for storage. Neem contains several aromatic principles that repel insects; for example, demonstration of the antifeedant properties of the neem kernel against the desert locust has generated tremendous interest in the insect-controlling properties of the plant. Spray applications of neem oil to rice plants was reported to inhibit the feeding responses of brown plant hopper and leaf folder, both rice pests. The application of neem oil to cowpea has demonstrated its effect in protecting the pulse from the infestation of bruchids. Farmers' experiments have indicated it as promising material for increasing biological nitrogen fixation in wetland paddy fields.

There are many other plants that our farmers use as insecticides and pesticides, building on their knowledge that has evolved over centuries. Chrysanthemum is a cosmopolitan genus comprising three hundred species of herbs and undershrubs, of which only a few have the insecticidal property. Pyrethrum, the relevant substance present in the flower, is one of the safest insecticides known; it has low mammalian toxicity and an instantaneous knockdown effect. Its repellent action toward insects even in very low concentrations makes it useful for the preservation of food grains and preparation of insect-resistant packaging. Indian farmers plant it around their fields to provide protection to other plants. The oil cake left after extraction of oil from the seeds of *pangamia glabra* trees also serves as a very potent pest control agent when added to the soil. The tree kasorka (*strychnos muxuomica*), found in Malnad forests, grows up to sixty to ninety feet in height. The pesticidal properties present in its seeds and leaves are known to our farmers, who have been using its leaves, barks, and twigs for pesticidal purposes since time immemorial.

Indian farmers depend on biodiversity for green and organic manure for their fields as well as fodder for their livestock. Soil is often described as consisting of solid particles, water, gaseous elements, humus, and raw organic matter. Organic matter serves as a nutrient store from which the nutrients are slowly released into the soil and made available to plants. Trees, shrubs, cover crops, grain, legumes, grasses, weeds, ferns, and algae all provide green manure; green manure crops contribute thirty to sixty kilograms of nitrogen per hectare annually. The cumulative effects of the continued use of green manure are important not only in terms of nitrogen supply but also with regard to soil organic matter and microelements.

Deep-rooted green manure crops in rotation can help recover nutrients leached to the subsoil. Similarly, there is a balance maintained between the animal population and fodder availability in the ecosystem. Trees, including fodder trees, are grown in combination with agricultural crops useful for producing fodder for livestock.

Long before the introduction of chemical fertilizers in Indian agriculture, oilseed cakes, particularly those of peanut (*arachis hypogaea*), castor (*ricinus coimmunis*), and mohua (*bassia latifolid*), were used as a source of plant nutrients. Scientists have reported on the value of the seed, bark, and leaf of karanji (*pongamia glabtra*) as manure in the Deccan region. Other plants that contribute to green manure are thangadi (*Cassia anriculosts*), yekka (*calitropics gigantea*), neem (*azadirachta indica*), the creeper uganishambu (*pettsonia spp*), and wild indigo (*tephrosia purpurea*). Other kinds of green manure collected from the jungle are portia (*thespesia populnuraa*), four o'clock plant (*mirabiulis jalepe*), and all pilli persara (*phaseolus aconitifilius*). Crops that contribute to green manure are pulses, for example, green gram, horse gram, black gram, cowpeas, and other legumes.

As for fodder for the animals, the tree *prosopis cineraria* is a most useful plant in dry parts of the country. There is a popular saying among farmers that death will not visit a man even during a famine if he has a *prosopis cineraria,* a goat, and a camel, because the three together will sustain him even under the most trying conditions. In wetland cultivation, it has been observed that green manure directly enhances soil conditions, whereas in dryland areas, fodder from animal dung is a rich source of manure. Local tall varieties of rice and millet are also an important source of fodder, which in turn returns to the soil as farmyard manure.

Thus, farmers' traditional knowledge of biodiversity use helps in increasing yields and protecting the environment by providing internal inputs as substitutes to economically expensive and environmentally destructive agrochemicals.

The conflict and contest between the two systems—one based on eco-imperialism, bioimperialism, and ecoapartheid, the other based on earth democracy and biodemocracy, will intensify over the next decade. People will have to strengthen their defenses to protect their local, living economies through local, living democracy.

Notes

1. *Business Line,* February 5, 2011.

2. *Outlook,* January 31, 2011.

3. Paul Collier, *The Plundered Planet* (Oxford: Oxford University Press, 2010).

4. Quoted in Peter Wahl, "Speculation Undermines the Right to Food," 2008.

5. Agribusiness Accountability Initiative, "Time to Act on Food Price Speculation," April 20, 2008, www.agribusinessaccountability.org.

6. Frederick Kaufman, "The Food Bubble," *Harper's,* July 2010.

7. Data from Navdanya/RFSTE field study "Skyrocketing Prices," January 2011.

8. Angus Wright, *The Death of Ramon Gonzalez: The Modern Agricultural Dilemma* (Austin: University of Texas Press, 1990).

9. ETC Group, *The New Biomassters: Synthetic Biology and the Next Assault on Biodiversity and Livelihoods* (2010).

10. Ibid.

5

Monocultures of the Mind

The "Disappeared" Knowledge Systems

In Argentina, when the dominant political system faces dissent, it responds by making the dissidents disappear. The *desaparecidos,* or the disappeared dissidents, share the fate of local knowledge systems throughout the world, which have been conquered through the politics of disappearance, not the politics of debate and dialogue.

The disappearance of local knowledge through its interaction with the dominant Western knowledge takes place at many levels, through many steps. First, local knowledge is made to disappear by simply not seeing it, by negating its very existence. This is very easy in the distant gaze of the globalizing dominant system. The Western systems of knowledge have generally been viewed as universal. However, the dominant system is also a local system, with its social basis in a particular culture, class, and gender. It is not universal in an epistemological sense. It is merely the globalized version of a very local and parochial tradition. Emerging from a dominating and colonizing culture, modern knowledge systems are themselves colonizing.

The knowledge and power nexus is inherent in the dominant system because, as a conceptual framework, it is associated with a set of values based on power that emerged with the rise of commercial capitalism. It generates inequalities and domination by the way such knowledge is generated and structured, the way it is legitimized and alternatives are delegitimized, and by the way in which such knowledge transforms nature and society. Power is also built into the perspective that views the dominant system not as a globalized local tradition but as a universal tradition, inherently superior to local systems. However, the dominant system is also the product of a particular culture. As Harding observes: "We can now discern the effects of these cultural markings in the discrepancies between the methods of knowing and the interpretations of the world provided by the creators of modern western culture and those characteristics of the rest of us. Western culture's favourite beliefs mirror in sometimes clear and sometimes distorting ways not the world as it is or as we might want it to be, but the social projects of their historically identifiable creators."[1] The universal/local dichotomy is misplaced when applied to the Western and indigenous traditions of knowledge, because the Western is a local tradition that has been spread worldwide through intellectual colonization. The universal would spread in openness. The globalizing local spreads by violence and misrepresentation. The first level of violence unleashed on local systems of knowledge is to not see them as knowledge. This invisibility is the first reason why local systems collapse without trial and test when confronted with the knowledge of the dominant West. The distance itself removes local systems from perception. When local knowledge does appear in the field of the globalizing vision, it is made to disappear by denying it the status of a systematic knowledge, and assigning it the adjectives "primitive" and "unscientific." Correspondingly, the Western system is assumed to be uniquely "scientific" and universal. The designation "scientific" for the modern systems and "unscientific" for the traditional knowledge systems has, however, less to do with knowledge and more to do with power. The models of modern science that have encouraged these perceptions were derived less from familiarity with actual scientific practice and more from idealized versions that gave science a special epistemological status. Positivism, verificationism, falsificationism were all based on the assumption that unlike traditional, local beliefs of the world, which are socially constructed, modern scientific knowledge was thought to be determined without social mediation. Scientists, in accordance with an abstract scientific method, were viewed as putting forward statements corresponding to the realities of a directly observable world. The theoretical concepts in their discourse were in principle seen as reducible to directly verifiable observational claims. New trends in the philosophy and sociology of science challenged the positivist assumptions but did not challenge the assumed superiority of Western systems. Thus, Kuhn, who has shown that science is not nearly as open as is popularly thought, and is the result of the commitment of a specialist community of scientists to presupposed metaphors and paradigms that determine the meaning of constituent terms and concepts, still holds that modern "paradigmatic" knowledge is superior to preparadigmatic knowledge, which represents a kind of primitive state of knowing.[2]

Horton, who has argued against the prevailing view of dominant knowledge, still speaks of the "superior cognitive powers" of the modes of thought of the modern scientific culture, which constitute forms of explanation, prediction, and control of a power unrivaled in any time and place. This cognitive superiority, in his view, arises from the "openness" of modern scientific thinking and the "closure" of traditional knowledge. As he interprets it, "In traditional cultures there is no developed awareness of alternatives to the established body of theoretical levels, whereas in the scientifically oriented cultures, such an awareness is highly developed."[3]

However, the historical experience of non-Western culture suggests that it is the Western systems of knowledge that are blind to alternatives. The "scientific" label assigns a kind of sacredness or social immunity to the Western system. By elevating itself above society and other knowledge systems and by simultaneously excluding other knowledge systems from the domain of reliable and systematic knowledge, the dominant system creates its exclusive monopoly. Paradoxically, it is the knowledge systems that are considered most open that are, in reality, closed to scrutiny and evaluation. Modern Western science is not to be evaluated; it is merely to be accepted. As Sandra Harding has said: "Neither God nor tradition is privileged with the same credibility as scientific rationality in modern cultures. . . . The project that science's sacredness makes taboo is the examination of science in just the ways any other institution or set of social practices can be examined."[4]

The Cracks of Fragmentation

Over and above rendering local knowledge invisible by declaring it nonexistent or illegitimate, the dominant system also makes alternatives disappear by erasing and destroying the reality that they attempt to represent. The fragmented linearity of the dominant knowledge disrupts the integrations between systems. Local knowledge slips through the cracks of fragmentation. It is eclipsed along with the world to which it relates. Dominant scientific knowledge thus breeds a monoculture of the mind by making space for local alternatives disappear, very much like monocultures of introduced plant varieties lead to the displacement and destruction of local diversity. Dominant knowledge also destroys the very *conditions* for alternatives to exist, very much like the introduction of monocultures destroy the very conditions for diverse species to exist.[5]

As metaphor, the monoculture of the mind is best illustrated in the knowledge and practice of forestry and agriculture. "Scientific" forestry and "scientific" agriculture split plants into artificially separate, non-overlapping domains on the basis of the separate commodity markets to which they supply raw materials and resources. In local knowledge systems, the plant world is not artificially separated between a forest supplying commercial wood and agricultural land supplying food commodities. The forest and the field are in ecological continuum, and activities in the forest contribute to the food needs of the local community, while agriculture itself is modeled on the ecology of the tropical forest. Some forest dwellers gather food directly from the forest, while many communities practice agriculture outside the forest but depend on the fertility of the forest for the fertility of agricultural land.

In the "scientific" system, which splits forestry from agriculture and reduces forestry to timber and wood supply, food is no longer a category related to forestry. The cognitive space that relates forestry to food production, either directly or through fertility links, is therefore erased with the split. Knowledge systems that have emerged from the food-giving capacities of the forest are therefore eclipsed and finally destroyed, both through neglect and aggression.

Most local knowledge systems have been based on the life-support capacities of tropical forests, not on their commercial timber value. These systems fall in the blind spot of a forestry perspective that is based exclusively on the commercial exploitation of forests. If some of the local uses can be commercialized, they are given the status of "minor products," with timber and wood being treated as the "major products" in forestry. The creation of fragmented categories thus blinkers out the entire spaces in which local knowledge exists, knowledge that is far closer to the life of the forest and more representative of its integrity and diversity. Dominant forestry science has no place for the

Local Knowledge Systems

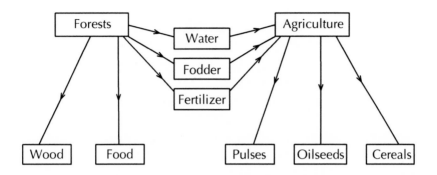

Dominant Knowledge Systems

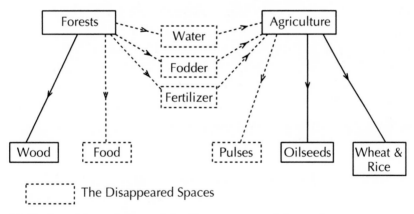

5.1. Dominant knowledge and the disappearance of alternatives

knowledge of the Hanunoo in the Philippines, who divide plants into sixteen hundred categories, of which trained botanists can distinguish only twelve hundred.[6] The knowledge base of the cropping systems based on 160 crops of the Lua tribe in Thailand is not counted as knowledge either by dominant forestry, which sees only commercial wood, or by dominant agriculture, which sees only chemically intensive agriculture. Food systems based on the forest, either directly or indirectly, are therefore nonexistent in the field of vision of a reductionist forestry and a reductionist agriculture, even though they have been and still are the sustenance base for many communities of the world. For example, the rainforests of Southeast Asia supply all the food needs of the Kayan, the Kenyah, the Punan Bah, the Penan, who gather food from the forest and practice swidden agriculture. The Tiruray people depend on the wild flora of the forests as a major source of food and other necessities.[7] The plant supplies are gathered mostly from the surrounding forest, and some 223 basic plant types are regularly exploited. The most important food items are mushrooms (*kulat*), ferns (*paku*), and the hearts of various plants (*ubot*) that include bamboo shoots, wild palms, and wild bananas. Twenty-five different varieties of fungi are eaten by the Kenyah and forty-three varieties are eaten by the Iban.[8] Sago, the staple of the Penan of Borneo, is the starch contained from the pith of a palm tree called the *Eugeissone utilis*. On New Guinea as a whole (Irian Jaya and Papua New Guinea together), 100,000 sago eaters produce 115,000 metric tons of sago each year.[9] Ethnobotanical work among India's many diverse tribes is also uncovering the deep, systematic knowledge of forests among them. The diversity of forest foods used in India emerges from this knowledge. In South India, a study conducted among the Soliga in the Belirangan hills of Karnataka shows that they use twenty-seven different varieties of leafy vegetables at different times of the year, and a variety of tubers, leaves, fruits, and roots are used for their medicinal properties by the tribes. A young illiterate Irula boy from a settlement near Kotagiri identified thirty-seven different varieties of plants and gave their Irula names and their different uses.[10]

In Madya Pradesh, although rice (*Oryza sativa*) and lesser millets (*Panicum miliaceum, Eleusine coracana,* and *Paspalum scrobiculatum*) form the staple diet of the tribes, almost all of them supplement it with seeds, grains, roots, rhizomes, leaves, and fruits of numerous wild plants that abound in the forests. Grigson noted that famine has never been a problem in Bastar as the tribes have always been able to draw half of their food from the innumerable edible forest products. Tiwari prepared a detailed list of wild plant species eaten by the tribes in Madhya Pradesh. He has listed 165 trees, shrubs, and climbers. Of these, the first category contains a list of 31 plants whose seeds are roasted and eaten. There are 19 plants whose roots and tubers are eaten after baking, boiling, or processing; there are 17 whose juice is taken fresh or after fermenting; 25 whose leaves are eaten as vegetables; and 10 whose petals are cooked as vegetables. There are 63 plants whose fruits are eaten raw, ripe, roasted, or pickled; there are 5 species of *Ficus* that provide figs for the forest dwellers. The fruits of the thorny shrub, *Pithcellobium dulce* (*Inga dulcis*), also called jungle jalebi, are favorites with the tribes. The sepals of mohwa are greedily eaten and also fermented for liquor. *Morus alba,* the mulberry, provides fruit for both people and birds. Besides, the ber (*Zizyphus mauritania* and *Z oenoplia*) provides delicious fruits, and has been eaten by jungle dwellers from the Mesolithic period onward.[11]

In nontribal areas, too, forests provide food and livelihood through critical inputs to agriculture, through soil and water conservation, and through inputs of fodder and organic fertilizer. Indigenous silvicultural practices are based on sustainable and renewable maximization of all the diverse forms and functions of forests and trees. This common silvicultural knowledge is passed on from generation to generation through participation in the processes of forest renewal and of drawing sustenance from the forest ecosystems.

In countries like India, the forest has been the source of fertility renewal of agriculture. The forest as a source of fodder and fertilizer has been a significant part of the agricultural ecosystem. In the Himalaya, the oak forests have been central to sustainability of agriculture. In the Western Ghats the "betta" lands have been central to the sustainability of the ancient spice gardens of pepper, cardamom, and areca nuts. Estimates show that over 50 percent of the total fodder supply for peasant communities in the Himalaya comes from forest sources, with forest trees supplying 20 percent.[12] In Dehra Dun, 57 percent of the annual fodder supply comes from the forests.[13] Besides fodder inputs, forests also make an important contribution to hill farming in the use of plant biomass as bedding for animals. Forests are the principal source of fallen dry leaf litter and lopped green foliage of trees and herbaceous species that are used for animal bedding and composting. Forest biomass, when mixed with animal dung, forms the principal source of soil nutrients for hill agriculture. On one estimate, 2.4 metric tons of litter and manure are used per hectare of cultivated land annually.[14] As this input declines, agricultural yields also go down.

The diverse knowledge systems that have evolved with the diverse uses of the forest for food and agriculture were eclipsed with the introduction of "scientific" forestry, which treated the forest only as a source of industrial and commercial timber. The linkages between forests and agriculture were broken, and the function of the forest as a source of food was no longer perceived.

When the West colonized Asia, it colonized the forests. It brought with it the ideas of nature and culture as derived from the model of the industrial factory. The forest was no longer viewed as having a value itself, in all its diversity. Its value was reduced to the value of commercially exploitable industrial timber. Having depleted their forests at home, European countries started the destruction of Asia's forests. England searched in the colonies for timber for its navy because the oak forests in England were depleted.

The military needs for Indian teak led immediately to a proclamation that wrested the right in teak trees from the local government and vested it in the East India Company. It was only after more than half a century of uncontrolled destruction of forests by British commercial interests that an attempt was made to control exploitation. In 1865, the first Indian Forest Act (VII of 1865) was passed by the supreme Legislative Council, which authorized the government to appropriate forests from the local people and manage them as reserved forests.

The introduction of this legislation marks the beginning of what state and industrial interests have called "scientific" management. However, for the indigenous people, it amounted to the beginning of the destruction of forests and erosion of people's rights to use the forests. The forest, however, is not merely a timber mine; it is also the source of food for local communities, and with the use of forests for food and for agriculture are related diverse knowledge systems. The separation of forestry from agriculture and the exclusive focus on wood production as the objective of forestry led to the creation of a one-dimensional forestry paradigm and the destruction of the multidimensional knowledge systems of forest dwellers and forest users.

"Scientific forestry" was the false universalization of a local tradition of forestry emerging from narrow commercial interests that viewed the forest only in terms of commercially valuable wood. It first reduced the value of diversity of life in the forest to the value of a few commercially valuable species, and further reduced the value of these species to the value of their dead product—wood. The reductionism of the scientific forestry paradigm created by commercial industrial interests violates both the integrity of the forests and the integrity of forest cultures that need the forests in their diversity to satisfy their needs for food, fiber, and shelter.

The principles of scientific forest management lead to the destruction of the tropical forest ecosystem because they are based on the objective of modeling the diversity of the living forest on the uniformity of the assembly line. Instead of society being modeled on the forest, as is the case for forest cultures, the forest is modeled on the factory. The system of "scientific management" as it has been practiced over a century is thus a system of tropical deforestation that transforms the forest from a renewable to a nonrenewable resource. Tropical timber exploitation thus becomes like mining, and tropical forests become a timber mine. According to an FAO estimate, at current rates of exploitation, the forests of tropical Asia will be totally exhausted by the turn of the century.

When modeled on the factory and used as a timber mine, the tropical forest becomes a nonrenewable resource. Tropical peoples also become dispensable. In place of cultural and biological pluralism, the factory produces nonsustainable monocultures in nature and society. There is no place for the small, no value for the insignificant. Organic diversity gives way to fragmented atomism and uniformity. The diversity must be weeded out, and the uniform monocultures—of plants and people—must now be externally managed because they are no longer self-regulated and self-governed. Those that do not fit into the uniformity must be declared unfit. Symbiosis must give way to competition, domination, and dispensability. There is no survival possible for the forest or its people when they become feedstock for industry. The survival of the tropical forests depends on the survival of human societies modeled on the principles of the forest. These lessons for survival do not come from texts of "scientific forestry." They lie hidden in the lives and beliefs of the forest peoples of the world.

There are in Asia today two paradigms of forestry—one life enhancing, the other life destroying. The life-enhancing paradigm emerges from the forest and the forest communities—the life-destroying one from the market. The life-enhancing paradigm creates a sustainable, renewable forest system, supporting and renewing food and water systems. The maintenance of conditions for renewability is the primary management objective of the former. The maximizing of profits through commercial extraction is the primary management objective of the latter. Since maximizing profits is consequent upon the destruction of conditions of renewability, the two paradigms are cognitively and ecologically incommensurate. Today in the forests of Asia the two paradigms are struggling against each other. This struggle is very clear in the two slogans on the utility of the Himalayan forests, one emanating from the ecological concepts of Garhwali women, the other from the sectoral concepts of those associated with trade in forest products. When Chipko became an ecological movement in 1977 in Adwani, the spirit of local science was captured in the slogan

> *What do the forests bear?*
> *Soil, water, and pure air.*

This was the response to the commonly accepted slogan of the dominant science:

> *What do the forests bear?*
> *Profit on resin and timber.*

The insight in these slogans represented a cognitive shift in the evolution of Chipko. The movement was transformed qualitatively from being based merely on conflicts over resources to involving conflicts over scientific perceptions and philosophical approaches to nature. This transformation also created that element of scientific knowledge that has allowed Chipko to reproduce itself in different ecological and cultural contexts. The slogan has become the scientific and philosophical message of the movement, and has laid the foundations of an alternative forestry science oriented to the public interest and ecological in nature. The commercial interest has the primary objective of maximizing exchange value through the extraction of commercially valuable species. Forest ecosystems are therefore reduced to the timber of commercially valuable species.

"Scientific forestry" in its present form is a reductionist system of knowledge that ignores the complex relationships within the forest community and between plant life and other resources like soil and water. Its pattern of resource utilization is based on increasing "productivity" on these reductionist foundations. By ignoring the system's linkages within the forest ecosystem, this pattern of resource use generates instabilities in the ecosystem and leads to a counterproductive use of natural resources at the ecosystem level. The destruction of the forest ecosystem and the multiple functions of forest resources in turn hurts the economic interests of those sections of society that depend on the diverse resource functions of the forests for their survival. These include soil and water stabilization and the provision of food, fodder, fuel, fertilizer, and so on.

Forest movements like Chipko are simultaneously a critique of reductionist "scientific" forestry and an articulation of a framework for an alternative forestry science that is ecological and can safeguard the public interest. In this alternative forestry science, forest resources are not viewed as isolated from other resources of the ecosystem. Nor is the economic value of a forest reduced to the commercial value of timber.

"Productivity," "yield," and "economic value" are defined for the integrated ecosystem and for multipurpose utilization. Their meaning and measure are therefore entirely different from the meaning and measure employed in reductionist forestry. Just as in the shift from Newtonian to Einsteinian physics the meaning of "mass" changed from a velocity-independent to a velocity-dependent term, in a shift from reductionist forestry to ecological forestry, all scientific terms are changed from ecosystem-independent to ecosystem-dependent ones. Thus, while for tribes and other forest communities a complex ecosystem is productive in terms of herbs, tubers, fiber, gene pool, and so on, for the forester, these components of the forest's ecosystem are useless, unproductive, dispensable.

The Chipko and Appiko movements are movements of agricultural communities against the destruction of the forests that support agriculture. The timber blockades of the Penan and other tribes of Sarawak are struggles of forest peoples against systems of forest management that destroy the forest and its people. According to the tribes:

This is the land of our forefathers, and their forefathers before them. If we don't do something now to protect the little that is left, there will be nothing for our children. Our forests are mowed down, the hills are levelled, the sacred graves of our ancestors have been desecrated, our waters and our streams are contaminated, our plant life is destroyed, and the forest animals are killed or have run away. What else can we do now but to make our protests heard, so that something can be done to help us?

AVEKMATAIAME MANEU MAPAT (until we die we will block this road).[15]

The Destruction of Diversity as "Weeds"

The destruction of biological diversity is intrinsic to the very manner in which the reductionist forestry paradigm conceives of the forest. The forest is defined as "normal" according to the objective of managing the forest for maximizing production of marketable timber. Since the natural tropical forest is characterized by richness in diversity, including the diversity of nonmarketable, nonindustrial species, the "scientific forestry" paradigm declares the natural forest as "abnormal." In Schlich's words, forest management implies that "the abnormal conditions are to be removed,"[16] and according to Troup, "The attainment of the normal forest from the abnormal condition of our existing natural forest, involves a certain temporary sacrifice. Generally speaking, the more rapid the change to the normal state, the greater the sacrifice; for example, the normal forests can be attained in one rotation by a series of clear fellings with artificial regeneration, but in an irregular, uneven-aged forest this means the sacrifice of much young growth which may be unsaleable. The question of minimising the sacrifice involved in introducing order out of chaos is likely to exercise our minds considerably in connection with forest management."[17]

The natural forest, in its diversity, is thus seen as "chaos." The man-made forest is "order." "Scientific" management of forests therefore has a clear antinature bias and a bias for industrial and commercial objectives, for which the natural forest must be sacrificed. Diversity thus gives way to uniformity of even-aged, single-species stands, and this uniformity is the ideal of the normal forestry toward which all silvicultural systems aim. The destruction and dispensability of diversity is intrinsic to forest management guided by the objective of maximizing commercial wood production, which sees noncommercial parts and relationships of a forest ecosystem as valueless—as weeds to be destroyed. Nature's wealth characterized by diversity is destroyed to create commercial wealth characterized by uniformity.

In biological terms, tropical forests are the most productive biological systems on our planet. A large biomass is generally characteristic of tropical forests. The quantities of wood especially are large in tropical forests and average about 300 tons per hectare, compared with about 150 tons per hectare for temperate forests. However, in reductionist commercial forestry, the overall productivity is not important, and nor are the functions of tropical forests in the survival of tropical peoples. It looks only for the industrially useful species that can be profitably marketed and measures productivity in terms of industrial and commercial biomass alone. It sees the rest as waste and weeds. As Bethel, an international forestry consultant, states, referring to the large biomass typical of the forests of the humid tropics: "It must be said that from a standpoint of industrial material supply, this is relatively unimportant. The important question is how much of this biomass represents trees and parts of trees of *preferred species that can be profitably marketed.* . . . By today's utilisation standards, *most of the trees, in these humid tropical forests are, from an industrial materials standpoint, clearly weeds.*"[18]

The industrial materials standpoint is the capitalist reductionist forestry that splits the living diversity and democracy of the forest into commercially valuable dead wood and destroys the rest as "weeds" and "waste." This "waste," however, is the wealth of biomass that maintains nature's water and nutrient cycles and satisfies the needs for food, fuel, fodder, fertilizer, fiber, and medicine of agricultural communities.

Just as "scientific" forestry excludes the food-producing functions of the forest and destroys the forest diversity as "weeds," "scientific" agriculture too destroys species that are useful as food, even though they may not be useful on the market. The Green Revolution has displaced not just seed varieties but entire crops in the third world. Just as people's seeds were declared "primitive" and "inferior" by the Green Revolution ideology, food crops were declared "marginal," "inferior," and "coarse grained."

Only a biased agricultural science rooted in capitalist patriarchy could declare nutritious crops like *ragi* and *jowar* inferior. Peasant women know the nutrition needs of their families and the nutritive content of the crops they grow. Among food crops they prefer those with maximum nutrition to those with a value in the market. What have usually been called "marginal crops" or "coarse grains" are nature's most productive crops in terms of nutrition. That is why women in Garhwal continue to cultivate *mandua* and women in Karnataka cultivate ragi in spite of all attempts by state policy to shift to cash crops and commercial food grains, to which all financial incentives of agricultural "development" are tied. Table 5.1 illustrates how what the Green Revolution has declared "inferior" grains are actually superior in nutritive content to the so-called superior grains, rice, and wheat. A woman in a Himalaya village once told me, "Without our mandua and jhangora, we could not labor as we do. These grains are our source of health and strength."

Not being commercially useful, people's crops are treated as "weeds" and destroyed with poisons. The most extreme example of this destruction is that of *bathua,* an important green leafy vegetable with a very high nutritive value and rich in vitamin A, which grows as an associate of wheat. However, with intensive chemical fertilizer use, bathua becomes a major competitor of wheat and has been declared a "weed" that is killed with herbicides. Forty thousand children in India go blind each year for lack of vitamin A, and herbicides contribute to this tragedy by destroying the freely available sources of vitamin A. Thousands of rural women who make their living by basket- and mat-making with wild reeds and grasses are also losing their livelihoods because the increased use of herbicide is killing the reeds and grasses. The introduction of herbicide-resistant crops will increase herbicide use and thus increase the damage to economically and ecologically useful plant species. Herbicide resistance also excludes the possibility of rotational and mixed cropping, which are essential for a sustainable and ecologically balanced agriculture, since the other crops would be destroyed by the herbicide. U.S. estimates now show a loss of U.S. $4 billion per annum as a result of herbicide spraying. The destruction in India will be far greater because of higher plant diversity and the prevalence of diverse occupations based on plants and biomass.

Strategies for genetic engineering resistance that are destroying useful species of plants can also end up creating superweeds. There is an intimate relationship between weeds and crops, especially in the tropics, where weedy and cultivated varieties have genetically interacted over centuries and hybridize freely to produce new varieties. Genes for herbicides tolerance that genetic engineers are striving to introduce into crop plants may be transferred to neighboring weeds as a result of naturally occurring gene transfer.

Scarcities of locally useful plant varieties have been created because the dominant knowledge systems discount the value of local knowledge and declare locally useful plants to be "weeds." Since dominant knowledge is created from the perspective of increasing commercial output and responds only to values on the market, it cannot see the values assigned to plant diversity by local perceptions. Diversity is thus destroyed in plant communities and forest and peasant communities because in commercial logic it is not "useful." And as Cotton Mather, the famous witch hunter of Salem, Massachusetts, had stated, "What is not useful is vicious." It must therefore be destroyed. When what is useful and what is not is determined one-sidedly, all other systems of determining value are displaced.

Declaring a locally useful species a weed is another aspect of the politics of disappearance by which the space of local knowledge shrinks out of existence. The one-dimensional field of vision of the dominant system perceives only one value, based on the market, and it generates forestry and agricultural practices that aim at maximizing that value. Related to the destruction of diversity as valueless is the inevitability of monoculture as the only "productive" and "high-yield" system.

"Miracle Trees" and "Miracle Seeds"

The one-dimensional perspective of dominant knowledge is rooted in the intimate links of modern science with the market. As multidimensional integrations between agriculture and forestry at the local level are broken, new integrations between nonlocal markets and local resources are established. Since economic power is concentrated in these remote centers of exploitation, knowledge develops according to the linear logic of maximizing flow at the local level. The integrated forest and farm gives way to the separate spheres of forestry and agriculture. The diverse forest and agricultural ecosystems are reduced to "preferred" species by the selective annihilation of species that are not "useful" from the market perspective. Finally, the "preferred" species themselves have to be engineered and introduced on the basis of "preferred" traits. The natural, native diversity is displaced by introduced monocultures of trees and crops.

In forestry, as the paper and pulp industry rose in prominence, pulp species became the species "preferred" by the dominant knowledge system. Natural forests were clear-felled and replaced by monocultures of the exotic eucalyptus species that were good for pulping. However, "scientific" forestry did not project its practice as a particular response to the particular interest of the pulp industry. It projected its choice as based on universal and objective criteria of "fast growth" and "high yields." In the 1980s, when the concern about deforestation and its impact on local communities and ecological stability created the imperative for afforestation programs, the eucalyptus was proposed worldwide as a "miracle" tree. However, local communities everywhere seemed to think otherwise.

The main thrust of conservation movements like Chipko is that forests and trees are life-support systems and should be protected and regenerated for their biospheric functions. The monoculture mind, on the other hand, sees the natural forest and trees as "weeds" and converts even afforestation into deforestation and desertification. From life-support systems, trees are converted into green gold—all planting is motivated by the slogan "Money grows on trees." Whether they are schemes like social forestry or wasteland development, afforestation programs are conceived at the international level by "experts" whose philosophy of tree planting falls within the reductionist paradigm of producing wood for the market, not biomass for maintaining ecological cycles or satisfying local needs of food, fodder, and fertilizer. All official programs of afforestation, based on heavy funding and centralized decision making, act in two ways against the local knowledge systems—they destroy the forest as a diverse and self-producing system, and they destroy it as commons shared by a diversity of social groups, with even the smallest having rights, access, and entitlements.

"Social" Forestry and the "Miracle" Tree

Social forestry projects are a good example of single-species, single-commodity production plantations based on reductionist models that divorce forestry from agriculture and water management, and seeds from markets. A case study of World Bank–sponsored social forestry in the

Kolar district of Karnataka[19] is an illustration of reductionism and mal-development in forestry being extended to farmland. Decentered agroforestry, based on multiple species and private and common tree stands, has been India's age-old strategy for maintaining farm productivity in arid and semiarid zones. The honge, tamarind, jackfruit, and mango, the jola, gobli, kagli, and bamboo traditionally provided food and fodder, fertilizer and pesticide, fuel and small timber. The backyard of each rural home was a nursery, and each peasant a silviculturalist. The invisible, decentered agroforestry model was significant because the humblest of species and the smallest of people could participate in it, and with space for the small, everyone was involved in protecting and planting.

The reductionist mind took over tree planting with "social forestry." Plans were made in national and international capitals by people who could not know the purpose of the honge and the neem, and saw them as weeds. The experts decided that indigenous knowledge was worthless and "unscientific," and proceeded to destroy the diversity of indigenous species by replacing them with row after row of eucalyptus seedlings in polythene bags in government nurseries. Nature's locally available seeds were laid waste; people's locally available knowledge and energies were laid waste. With imported seeds and expertise came the import of loans and debt and the export of wood, soils, and people. Trees, as a living resource maintaining the life of the soil and water and of local people, were replaced by trees whose dead wood went straight to a pulp factory hundreds of miles away. The smallest farm became a supplier of raw material to industry and ceased to be a supplier of food to local people. Local work linking the trees to the crops disappeared and was replaced by the work of brokers and middlemen who brought the eucalyptus trees on behalf of industry. Industrialists, foresters, and bureaucrats loved the eucalyptus because it grows straight and is excellent pulpwood, unlike the honge, which shelters the soil with its profuse branches and dense canopy and whose real worth is as a living tree on a farm.

The honge could be nature's idea of the perfect tree for arid Karnataka. It has rapid growth of precisely those parts of the tree, the leaves and small branches, that go back to the earth, enriching and protecting it, conserving its moisture and fertility. The eucalyptus, on the other hand, when perceived ecologically, is unproductive, even negative, because this perception assesses the "growth" and "productivity" of trees in relation to the water cycle and its conservation, in relation to soil fertility, and in relation to human needs for food and food production. The eucalyptus has destroyed the water cycle in arid regions due to its high water demand and its failure to produce humus, which is nature's mechanism for conserving water.

Most indigenous species have a much higher biological productivity than the eucalyptus, when one considers water yields and water conservation. The nonwoody biomass of trees has never been assessed by forest measurements and quantification within the reductionist paradigm, yet it is this very biomass that functions in conserving water and building soils. It is little wonder that Garhwal women call a tree *dali*, or branch, because they see the productivity of the tree in terms of its nonwoody biomass that functions critically in hydrological and nutrient cycles within the forest, and through green fertilizer and fodder in cropland.

Eucalyptus

The most powerful argument in favor of the expansion of eucalyptus is that it is faster growing than all indigenous alternatives. This is quite clearly untrue for ecozones where eucalyptus has had no productivity due to pest damage. It is also not true for zones with poor soils and poor water endowment, as the reports on yields make evident. Even where biotic and climatic factors are conducive to good growth, eucalyptus cannot compete with a number of indigenous fast-growing species. When overstated scientific claims about the growth rate of eucalyptus were being used to convert rich natural forests to eucalyptus monoculture plantations, on the grounds of the improvement of the productivity of the site, the central silviculturist and director of forestry research of the Forest Research Institute (FRI) had categorically stated that "some indigenous species are as fast growing as, and in some cases even more than, the much coveted Eucalyptus."[20] In justification he provided a long list of indigenous fast-growing species that had growth rates exceeding that of the eucalyptus, which under the best conditions is about ten cubic meters per hectare per year, and on average is about five cubic meters per hectare per year (table 5.2). Indigenous trees are those trees that are native to the Indian soil or are exotics that have been naturalized over thousands of years.

These data based on forest plantations do not include fast-growing farm tree species such as *Pongamia pinnata, Greivia optiva,* and others that have been cultivated for agricultural inputs to farms but have not been of interest in commercial forestry. In spite of being an incomplete list of fast-growing indigenous trees, the forest plantation data on yields adequately reveal that eucalyptus is among the slower-growing species even for woody biomass production.[21] The eucalyptus hybrid, the most dominantly planted eucalyptus species, has different growth rates at different ages and on different sites, as shown in table 5.3.

The points that emerge from tables 5.2 and 5.3 are:

1. In terms of yields measured as mean annual increment (MAI), eucalyptus is a slow producer of woody biomass even under very good soil conditions and water availability.
2. When the site is of poor quality such as eroded soils or barren land, eucalyptus yields are insignificant.
3. The growth rate of eucalyptus under the best conditions is not uniform for different age groups. It falls very drastically after five or six years.

Scientific evidence on biomass productivity does not support the claim that the eucalyptus is faster growing than other alternative species or that it grows well even on degraded lands. Under rain-fed conditions, the best yields achieved for eucalyptus have been ten tons per hectare per year. On the other hand: "According to Dr K S Rao and Dr K K Bokil (unpublished reports) one hectare of Prosopis yields 31 tons of bone dry firewood per year. At Vatva in Ahmedabad district, Gujarat state, annual production of firewood from Prosopis was recorded as 25 tons/ha/yr under rain-fed conditions."

A comparison of growth rate of ten species by the Gujarat Forest Department shows that the eucalyptus emerges at the bottom of the list. The eucalyptus, quite clearly, will not fill the gap in the demand of woody biomass more effectively than other faster-growing species, which are also better adapted to the Indian conditions.

Forests and trees have been producing various kinds of biomass satisfying diverse human needs. Modern forestry management, however, came as a response to the demands for woody biomass for commercial and industrial purposes. The growth rate of the species that is provided by modern forestry is, therefore, restricted in two ways. First, it is confined to the increment and growth of the trunk biomass alone. Even in this limited spectrum, eucalyptus ranks very low in terms of growth and biomass productivity. Second, it reduces the diversity of trees to one exotic species.

Human needs for biomass are, however, not restricted to the consumption and use of woody biomass alone. The maintenance of life-support systems is a function performed mainly by the crown biomass of trees. It is this component of trees that can contribute positively to the maintenance of the hydrological and nutrient cycles. It is also the most important source for the production of biomass for consumption as fuel, fodder, manure, fruits, and the like.[22] Social forestry, as distinct from commercial forestry, to which it is supposed to be a corrective, is, in principle, aimed at the maximization of the production of all types of useful biomass that improve ecological stability and satisfy diverse and basic biomass needs. The appropriate unit of assessment of growth and yields of different tree species for social forestry programs cannot be restricted to the woody biomass production for commercial use. It must, instead, be specific to the end use of biomass. The crisis in biomass for animal feed, quite evidently, cannot be overcome by planting trees that are fast growing from the perspective of the pulp industry but are absolutely unproductive as far as fodder requirements are concerned.[23]

The assessment of yields in social forestry must include the diverse types of biomass that provide inputs to the agroecosystem. When the objective for tree planting is the production of fodder or green fertilizer, it is relevant to measure crown biomass productivity. India, with its rich genetic diversity in plants and animals, is richly endowed with various types of fodder trees that have annual yields of crown biomass that is much higher than the total biomass produced by eucalyptus plantations, as indicated in table 5.4.[24]

An important biomass output of trees that is never assessed by foresters who look for timber and wood is the yield of seeds and fruits. Fruit trees such as jack, jaman, mango, tamarind, and so on have been important components of indigenous forms of social forestry as practiced over centuries in India. After a brief gestation period, fruit trees yield annual harvests of edible biomass on a sustainable and renewable basis.

Tamarind trees can yield fruits for over two centuries. Other trees, such as neem, pongamia, and sal, provide annual harvests of seeds that yield valuable nonedible oils. These diverse yields of biomass provide important sources of livelihood for millions of tribes or rural people. The coconut, for example, besides providing fruits and oil, offers leaves used in thatching huts and supports the large coir industry in the country. Since social forestry programs in their present form have been based solely on the knowledge of foresters who have been trained only to look for the woody biomass in the tree, these important high-yielding species of other forms of biomass have been totally ignored in these programs. Two species on which ancient farm forestry systems in arid zones have laid special stress are pongmia and tamarind. Both these trees are multidimensional producers of firewood, fertilizer, fodder, fruit, and oilseed. More significant, components of the crown biomass that are harvested from fruit and fodder trees leave the living tree standing to perform its essential ecological functions in soil and water conservation. In contrast, the biomass of the eucalyptus is useful only after the tree is felled.

Figures 5.2 and 5.3 describe the comparative biomass contribution of indigenous trees and eucalyptus. Afforestation strategies based dominantly on eucalyptus are not, therefore, the most effective mechanism for dealing with the serious biomass crisis facing the country. The benefits of eucalyptus have often been unduly exaggerated through the myth of its fast growth and high yields. The myth has become pervasive because of the unscientific and unjustified advertisement of the species. It has also been aided by the linear growth of eucalyptus in one dimension while most indigenous trees have broad crowns that grow in three dimensions.

The Green Revolution and "Miracle" Seeds

In agriculture, too, the monoculture mind creates the monoculture crop. The miracle of the new seeds has most often been communicated by the term *high-yielding varieties*. The HYV category is central to the Green Revolution paradigm. Unlike what the term suggests, there is no neutral or objective measure of "yield" on the basis of which the cropping systems founded on miracle seeds can be established to be higher yielding than the cropping systems they replace. It is now commonly accepted that even in the most rigorous of scientific disciplines such as physics, there are no neutral observational terms. All terms are theory laden.

The HYV category is similarly not a neutral observational concept. Its meaning and measure are determined by the theory and paradigm of the Green Revolution. And this meaning is not easily and directly translat-

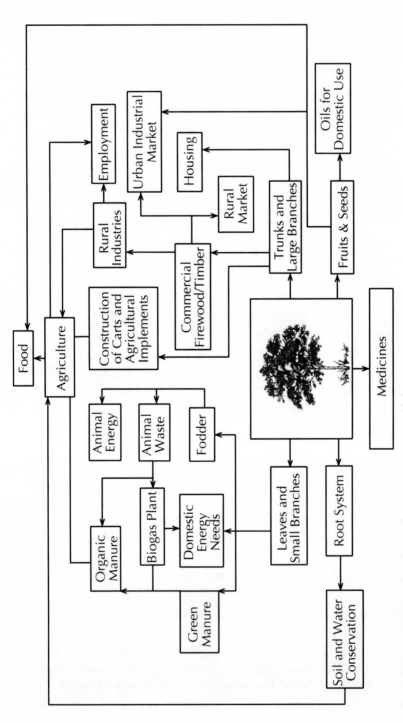

5.2. The contribution of traditional tree species to rural life-support systems

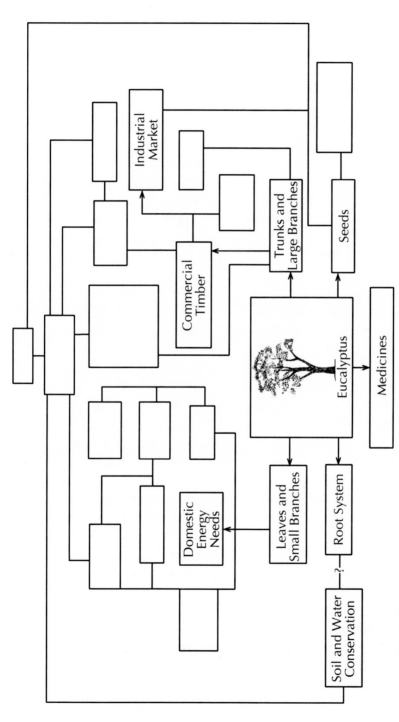

5.3. The comparative contribution of eucalyptus to rural life-support systems

able for comparison with the agricultural concept of indigenous farming systems for a number of reasons. The Green Revolution category of HYV is essentially a reductionist category that decontextualizes contextual properties of both the native and the new varieties. Through the process of decontextualization, costs and impacts are externalized and systemic comparison with alternatives is precluded.

Cropping systems, in general, involve an interaction between soil, water, and plant genetic resources. In indigenous agriculture, for example, cropping systems include a symbiotic relationship between soil, water, farm animals, and plants. Green Revolution agriculture replaces this integration at the level of the farm with the integration of inputs such as seeds and chemicals. The seed/chemical package sets up its own interactions with soils and water systems, which are, however, not taken into account in the assessment of yields.

Modern plant-breeding concepts like HYVs reduce farming systems to individual crops and parts of crops (figure 5.4). Crop components of one system are then measured with crop components of another. Since the Green Revolution strategy is aimed at increasing the output of a single component of a farm, at the cost of decreasing other components and increasing external inputs, such a partial comparison is by definition biased to make the new varieties "high yielding," although at the systems level they may not be.

Traditional farming systems are based on mixed and rotational cropping systems of cereals, pulses, and oilseeds, with diverse varieties of each crop, while the Green Revolution package is based on genetically uniform monocultures. No realistic assessments are ever made of the yield of the diverse-crop outputs in the mixed and rotational systems. Usually the yield of a single crop like wheat or maize is singled out and compared to yields of new varieties. Even if the yields of all the crops were included, it is difficult to convert a measure of pulse into an equivalent measure of wheat, for example, because in the diet and in the ecosystem, they have distinctive functions.

The protein value of pulses and the calorie value of cereals are both essential for a balanced diet but in different ways, and one cannot replace the other, as illustrated in table 5.1. Similarly, the nitrogen-fixing capacity of pulses is an invisible ecological contribution to the yield of associated cereals. The complex and diverse cropping systems based on indigenous

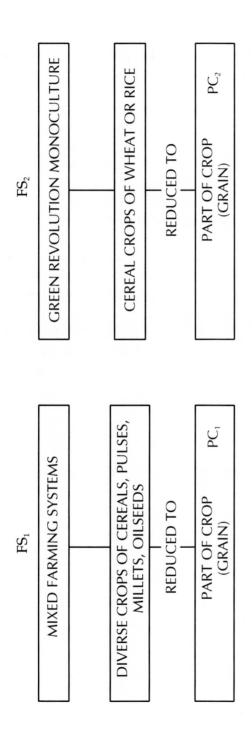

FS₁

MIXED FARMING SYSTEMS

DIVERSE CROPS OF CEREALS, PULSES, MILLETS, OILSEEDS

REDUCED TO

PART OF CROP (GRAIN) PC_1

FS₂

GREEN REVOLUTION MONOCULTURE

CEREAL CROPS OF WHEAT OR RICE

REDUCED TO

PART OF CROP (GRAIN) PC_2

➤ The real scientific comparison should be between two farming systems — FS_1 and FS_2, with the full range of inputs and outputs included.

➤ This would be the comparison if FS_2 were not given immunity from an ecological evaluation.

➤ In the Green Revolution strategy, a false comparison is made between PC_1 and PC_2.

➤ So while $PC_2 > PC_1$, generally $FS_1 > FS_2$

5.4. How the Green Revolution makes unfair comparisons

varieties are therefore not easy to compare to the simplified monocultures of HYV seeds. Such a comparison has to involve entire systems and cannot be reduced to a comparison of a fragment of the farm system. In traditional farming systems, production has also involved maintaining the conditions of productivity. The measurement of yields and productivity in the Green Revolution paradigm is divorced from seeing how the processes of increasing output affect the processes that sustain the conditions for agricultural production. While these reductionist categories of yield and productivity allow a higher destruction that affects future yields, they also exclude the perception of how the two systems differ dramatically in terms of inputs (figure 5.5).

The indigenous cropping systems are based only on internal organic inputs. Seeds come from the farm, soil fertility comes from the farm, and pest control is built into the crop mixtures. In the Green Revolution package, yields are intimately tied to purchased inputs of seeds, chemical fertilizers, pesticides, and petroleum, and to intensive and accurate irrigation. High yields are not intrinsic to the seeds but are a function of the availability of required inputs, which in turn have ecologically destructive impacts (figure 5.6).

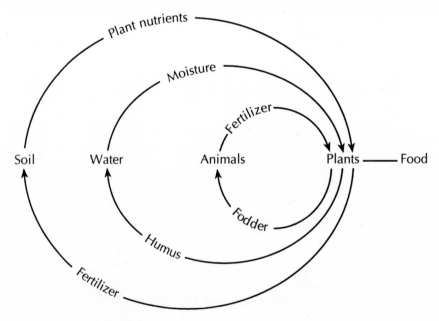

5.5. Internal input farming system

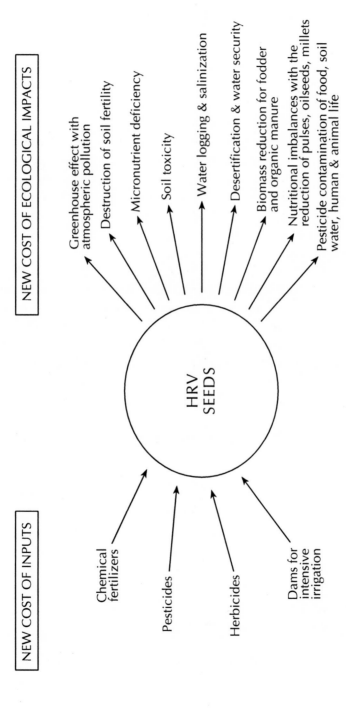

NEW COST OF INPUTS

NEW COST OF ECOLOGICAL IMPACTS

HRV SEEDS

Chemical fertilizers

Pesticides

Herbicides

Dams for intensive irrigation

Greenhouse effect with atmospheric pollution

Destruction of soil fertility

Micronutrient deficiency

Soil toxicity

Water logging & salinization

Desertification & water security

Biomass reduction for fodder and organic manure

Nutritional imbalances with the reduction of pulses, oilseeds, millets

Pesticide contamination of food, soil water, human & animal life

5.6. External input farming system

In the absence of additional inputs of fertilizers and irrigation, the new seeds perform worse than indigenous varieties. With the additional inputs, the gain in output is insignificant compared to the increase in inputs. The measurement of output is also biased by restricting it to the marketable part of crops. However, in a country like India, crops have traditionally been bred and cultivated to produce not just food for people but fodder for animals and organic fertilizer for soils. According to A. K. Yegna Iyengar, a leading authority on agriculture, "As an important fodder for cattle and in fact as the sole fodder in many tracts, the quantity of straw obtainable per acre is important in this country. Some varieties which are good yielders of grains suffer from the drawback of being low in respect to straw."[25] He illustrated the variation in the grain-straw ratio with yields from the Hebbal farm.

In the breeding strategy of the Green Revolution, multiple uses of plant biomass seem to have been consciously sacrificed for a single use, with nonsustainable consumption of fertilizer and water. The increase in marketable output of grain has been achieved at the cost of the decrease of biomass for animals and soils and the decrease of ecosystem productivity due to overuse of resources.

The increase in the production of grain for marketing was achieved in the Green Revolution strategy by reducing the biomass for internal use on the farm. This is explicit in a statement by Swaminathan:

> High yielding varieties of wheat and rice are high yielding because they can use efficiently larger quantities of nutrients and water than the earlier strains, which tended to lodge or fall down if grown in soils with good fertility. . . . They thus have a "harvest index" (i.e., the ratio of the economic yield to the total biological yield) which is more favourable to man. In other words, if a high yielding strain and an earlier tall variety of wheat both produce, under a given set of conditions, 1,000 kg of dry matter, the high yielding strain may partition this dry matter into 500 kg for grain and 500 kg for straw. The tall variety, on the other hand, may divert 300 kg for grain and 700 kg for straw.[26]

The reduction of outputs of biomass for straw production was probably not considered a serious cost since chemical fertilizers were viewed as a total substitute for organic manure, and mechanization was viewed as a substitute for animal power. According to one author, "It is believed that the 'Green Revolution' type of technological change permits higher grain production by changing the grain foliage ratio. . . . At a time there is urgency for increasing grain production, an engineering approach to altering the product mix on an individual plant may be advisable, even inevitable. This may be considered another type of survival technological change. It uses more resources, returns to which are perhaps unchanged (if not diminished)."[27] It was thus recognized that in terms of overall plant biomass, the Green Revolution varieties could even reduce the overall yields of crops and create scarcity in terms of output such as fodder.

Finally, there is now increasing evidence that indigenous varieties could also be high yielding, given the required inputs. Richaria has made a significant contribution to the recognition that peasants have been breeding high-yielding varieties for centuries. Richaria reports:

A recent varietal-cum-agronomic survey has shown that nearly 9% of the total varieties grown in U P fall under the category of high yielding types (3,705 kgs and above per hectare).

A farmer planting a rice variety called Mokdo of Bastar who adopted his own cultivation practices obtained about 3,700 to 4,700 kgs of paddy per hectare. Another rice grower of Dhamtari block (Raipur) with just a hectare of rice land, falling not in an uncommon category of farmers, told me that he obtains about 4,400 kgs of paddy per hectare from Chinnar variety, a renowned scented type, year after year with little fluctuations. He used FYM supplemented at times with a low dose of nitrogen fertilizers. For low lying area in Farasgaon Block (Bastar) a non-lodging tall rice variety Surja with bold grains and mildly scented rice may compete with Jaya in yield potential at lower doses of fertilization, according to a local grower who showed me his crop of Surja recently.

During my recent visit of the Bastar area in the middle of November 1975 when the harvesting of a new rice crop was in full swing in a locality, in one of the holdings of an adivasi cultivator, Baldeo of Bhatra tribe in village Dhikonga of Jugalpur block, I observed a field of Assam Chudi ready for harvest with which the adivasi cultivator has stood for crop competition. The cultivator has applied the fertilizer approximately equal to 50 kg/N ha and has used no plant protection measures. He expected a yield of about 5,000 kg/ha. These are good cases of applications of an intermediate technology for increasing rice production. The yields obtained by those farmers fall in or above the minimum limits set for high yields and these methods of cultivation deserve full attention.[28]

India is a Vavilov center, or center of genetic diversity of rice. Out of this amazing diversity, Indian peasants and tribals have selected and improved many indigenous high-yielding varieties. In South India, in semiarid tracts of the Deccan, yields went up to five thousand kilograms per hectare under tank and well irrigation. Under intensive manuring, they could go even higher. As Yegna Iyengar reports: "The possibility of obtaining phenomenal and almost unbelievably high yields of paddy in India has been established as the result of the crop competitions organised by the Central Government and conducted in all states. Thus even the lowest yield in these competitions has been about 5,300 lbs/acre, 6,200 lbs/acre in West Bengal, 6,100, 7,950,and 8,258 lbs/acre in Thirunelveli, 6,368 and 7,666 kg/ha in South Arcot, 11,000 lbs/acre in Coorg and 12,000 lbs/acre in Salem."[29]

The Green Revolution package was built on the displacement of genetic diversity at two levels. First, mixtures and rotation of diverse crops like wheat, maize, millets, pulses, and oilseeds were replaced by monocultures of wheat and rice. Second, the introduced wheat and rice varieties reproduced over a large scale as monocultures came from a very narrow genetic base, compared to the high genetic variability in the population of traditional wheat or rice plants. When "HYV" seeds replace native cropping systems, diversity is lost and is irreplaceable.

The destruction of diversity and the creation of uniformity simultaneously involve the destruction of stability and the creation of vulnerability. Local knowledge, on the other hand, focuses on multiple use of diversity. Rice is not just grain; it provides straw for thatching and mat-making, fodder for livestock, bran for fish ponds, husk for fuel. Local varieties of crops are selected to satisfy these multiple uses. The so-called HYV varieties increase grain production by decreasing all other outputs, increasing external inputs, and introducing ecologically destructive impacts.

Local knowledge systems have evolved tall varieties of rice and wheat to satisfy multiple needs. They have evolved sweet cassava varieties whose leaves are palatable as fresh greens. However, all dominant research on cassava has focused on breeding new varieties for tuber yields, with leaves that are unpalatable.

Ironically, breeding for a *reduction* in usefulness has been viewed as important in agriculture, because uses outside those that serve the market are not perceived and taken into account. The new ecological costs are also left out as "externalities," thus rendering an inefficient, wasteful system productive.

There is, moreover, a cultural bias that favors the modern system, a bias that becomes evident in the naming of plant varieties. The indigenous varieties, or landraces, evolved through both natural and human selection and produced and used by third world farmers worldwide, are called "promotive cultivar." Those varieties created by modern plant breeders in international agricultural research centers or by transnational seed corporations are called "advanced" or "elite."

Yet the only aspect in which the new varieties have really been "advanced" has been in their ecologically appropriate systems, not through test and evaluation but through the unscientific rejection of local knowledge as primitive and the false promise of "miracles"—of "miracle" trees and "miracle" seeds.

The Nonsustainability of Monocultures

The crucial characteristic of monocultures is that they do not merely displace alternatives, but they also destroy their own basis. They are neither tolerant of other systems, nor are they able to reproduce themselves sustainably. The uniformity of the "normal" forest that "scientific" forestry attempts to create becomes a prescription for nonsustainability.

The displacement of local forest knowledge by "scientific" forestry was simultaneously a displacement of the forest diversity and its substitution by uniform monocultures. Since the biological productivity of the forest is ecologically based on its diversity, the destruction of local knowledge, and with it of plant diversity, leads to a degradation of the forest and an undermining of its sustainability. The increase in productivity from the commercial point of view destroys productivity from the perspective of local communities. The uniformity of the managed forest is meant to generate "sustained yields." However, uniformity destroys the conditions of renewability of forest ecosystems and is ecologically nonsustainable.

In the commercial forestry paradigm, "sustainability" is a matter of supply to the market, not the reproduction of an ecosystem in its biological diversity or hydrological and climatic stability. As Schlich states, "Forest working plans regulate, according to time and locality, the management of forests in such a manner, that the objects of the industry are as full as possible realised."[30] Sustained yield management is aimed at producing "the best financial results, or the greatest volume, or the most suitable class of produce." If this could be ensured while maintaining the forest ecosystem, we would have sustainability in nature, not just short-term sustainability for market supplies of industrial and commercial wood. However, "sustained yields," as conceived in forestry management, are based on the assumption that the real forest, or the natural forest, is not a "normal" forest; it is an "abnormal" forest. When "normalcy" is determined by the demands of the market, the nonmarketable components of the natural forest ecosystem are seen as "abnormal" and are destroyed by prescriptions of forest working plans.

Uniformity in the forest is the demand of centralized markets and centralized industry. However, uniformity acts against nature's processes. The transformation of mixed natural forests into uniform monocultures allows the direct entry of tropical sun and rain, baking the forest soils dry in the heat, washing the soils off in the rain. Less humid conditions are the reason for rapid retrogression of forest regions. The recent fires of Kalimantan are largely related to the aridization caused by the conversion of rainforests into plantations of eucalyptus and acacias. Floods and drought are created where the tropical forest had earlier cushioned the discharge of water.

In tropical forests, selective felling of commercial species produces only small yields (5–25 cubic meters per hectare), whereas clear-felling might produce as much as 450 cubic meters per hectare. The nonsustainability of selection fellings is also borne out by the experience of PICOP, a joint venture set up in 1952 between the American firm International Paper Company, the world's largest paper producer, and the Andre Soriano

Corporation in the Philippines. The company takes only about 10 percent of the total volume of wood, roughly seventy-three cubic yards per acre of virgin forest. But the company's measurements of annual growth show that the second rotation will only yield thirty-seven cubic yards of useful wood per acre, half as much as the first cut, and not enough to keep the company's plywood, veneer, and sawmills functioning at a profitable level.

"Sustained yields" can be managed for PICOP by reducing the diameter for extraction. At present, the government allows PICOP to take out all trees larger than thirty-two inches in diameter, and a certain proportion of those that are twenty-four inches or more in diameter. If on the second rotation they could take out all trees bigger than twelve or sixteen inches around, they could sustain supplies for another rotation. Taking smaller trees on the second cut would not, of course, make the forest grow faster for a third, fourth, and fifth rotation.

PICOP's plantations have also failed. It had to replant thirty thousand acres of a variety of eucalyptus from Papua New Guinea that was attacked by pests. Its pine plantations of twenty-five thousand acres have also failed. At $400 per acre, that was a $10 million mistake.

Angel Alcala, professor of biology at Siliman University in the Philippines, observes that selective logging is good in theory, but it does not really work. "With selective logging, you are supposed to take only a few trees and leave the rest to grow, so you can return later and take some more, without destroying the forest. This is supposed to be a sustainable system. But here, although they use the phrase selective logging, there is only one harvest, a big one. After that no more."[31]

One study found that 14 percent of a logging area is cleared for roads and another 27 percent for skidder trucks. Thus, more than 40 percent of a concession can be stripped of protective vegetation and become highly liable to erosion. It can be as high as 60 percent.[32]

In dipterocarp forests, with an average of fifty-eight trees per acre, for every ten that are deliberately felled, thirteen more are broken or damaged. Selective loggers damage more trees than they harvest. In one Malaysian dipterocarp forest, only 10 percent of the trees were harvested; 55 percent were destroyed or severely damaged. Only 33 percent were unharmed. In Indonesia, according to the manager of Georgia-Pacific, the company damages or destroys more than three times as many as it deliberately harvests.[33]

According to the UNESCO report on tropical forest ecosystems, not many forests are rich enough to allow true selective working—the removal of each tree (of desirable species) as soon as it reaches commercial size. Not only does each tree cause considerable damage when it falls, but the heavy logging equipment needed causes further damage. To sum up, true selective felling is impracticable regardless of the structure, composition, and dynamism of the original stands.

This paradigm, which destroys the diversity of the forest community, either by clear-felling or selective felling, simultaneously destroys the very *conditions* for the renewal of the forest community. While species diversity is what makes the tropical forest biologically rich and sustainable, this same diversity allows density of individual species. The reductionist paradigm thus converts a biologically rich system into an impoverished resource and hence a nonrenewable one. Thus, while the annual biological production of tropical broadleaved forest is 300 tons per hectare compared to 150 tons per hectare, the annual production of commercial wood is only 0.14 cubic meters per hectare on the average in tropical forests, compared to 1.08 cubic meters. In tropical Asia, commercial production is 0.39 cubic meters per hectare due to the richness in diversity of commercial species of the dipterocarp forests.[34]

In the dominant system, financial survival strategies determine the concept of "sustained yield," which are in total violation of the principles of sustaining biological productivity. Sustained yields based on continually reducing exploitable diameter classes leads to biological suicide and a total destruction of forests.

Fahser reports how a forestry project in Brazil, aimed at "self-help" and satisfying basic needs, destroyed both the forests and the communities it was intended to improve:

> With the building up of the first Faculty of Forestry Science and the imparting of modern forestry knowledge, a milestone was actually reached in the forests of Brazil. A greater knowledge of economics encouraged trained people towards new approaches; the natural forest with its many species was replaced by huge timber plantations of fir and eucalyptus; weak and unreliable human workers were replaced by powerful timber harvesting machinery; the hitherto untouched coastal mountain ranges

were conquered, using rope cranes as an elegant means of transport.

Since forestry development aid began, afforestation in Parana has dropped from about 40% to its present level of 8%. Transformation into steppe, erosion and periodical flooding are on the increase. Our highly qualified Brazilian counterparts are now shifting their interest to the Amazon regions of the north where there are still plenty of forests and where they are "managing" cellulose timber plantations (e.g., of Gmelina arborea) with rotation periods of only six years.

What happened to the population during the roughly 20 years period of the project, to those people whose basic needs were to be satisfied and who were to be given aid so that they could help themselves? Parana is now largely cleared of forest and full of mechanised agriculture. Most Indios and many immigrants who lived there at subsistence level or as small farmers have silently disappeared, become impoverished and collected in the slums (favelas) in the vicinity of the cities. In forestry the capital-intensive unit on the mechanisation pattern of north America and Scandinavia is now dominant. Only a few experts and a few wage-earners are still needed for peak work periods.[35]

Where the local knowledge is not totally extinct, communities resist the ecological destruction of introduced monocultures. "Greening" with eucalyptus works against nature and its cycles, and it is being resisted by communities that depend on the stability of nature's cycles to provide sustenance in the form of food and water. The eucalyptus guzzles nutrients and water and, in the specific conditions of low rainfall zones, gives nothing back but terpenes to the soil. These inhibit the growth of other plants and are toxic to soil organisms that are responsible for building soil fertility and improving soil structure. The eucalyptus certainly increased cash and commodity flows, but it resulted in a disastrous interruption of organic matter and water flows within the local ecosystem. Its proponents failed to calculate the costs in terms of the destruction of life in the soil, the depletion of water resources, and the scarcity of food and fodder that eucalyptus cultivation creates. Nor did they, while trying to shorten rotations for harvesting, see that tamarind, jackfruit, and honge have very short rotations of one year in which the biomass harvested is far higher than that of eucalyptus, which they nevertheless declared a "miracle" tree. The crux of the matter is that fruit production was never the concern of forestry in the reductionist paradigm—it focused on wood, and wood for the market, alone. Eucalyptus as an exotic, introduced in total disregard of its ecological appropriateness, has thus become an exemplar of antilife afforestation.[36]

People everywhere have resisted the expansion of eucalyptus because of its destruction of water, soil, and food systems. On August 10, 1983, the small peasants of Barha and Holahalli villages in Tumkur district (Karnataka) marched en masse to the forestry nursery and pulled out millions of eucalyptus seedlings, planting tamarind and mango seeds in their place. This gesture of protest, for which they were arrested, spoke out against the virtual planned destruction of soil and water systems by eucalyptus cultivation. It also challenged the domination of a forestry science that had reduced all species to one (the eucalyptus), all needs to one (that of the pulp industry), and all knowledge to one (that of the World Bank and forest officials). It challenged the myth of the miracle tree: tamarind and mango are symbols of the energies of nature and of local people, of the links between these seeds and the soil, and of the needs that these trees—and others like them—satisfy in keeping the earth and the people alive. Forestry for food—food for the soil, for farm animals, for people—all women's and peasants' struggles revolve around this theme, whether in Garhwal or Karnataka, in the Santhal Perganas or Chattisgarh, in reserved forests, farmlands, or commons. In June 1988, in protest against eucalyptus planting, villagers in northern Thailand burned down eucalyptus nurseries at a forestry station.

The destruction of diversity in agriculture has also been a source of nonsustainability. The "miracle" varieties displaced the traditionally grown crops, and through the erosion of diversity, the new seeds became a mechanism for introducing and fostering pests. Indigenous varieties, or landraces, are resistant to locally occurring pests and diseases. Even if certain diseases occur, some of the strains may be susceptible, while others will have the resistance to survive. Crop rotations also help in pest control. Since many pests are specific to particular plants, planting crops in different seasons and different years causes large reductions in pest populations. On the other hand, planting the same crop over large areas year after year encourages pest buildups. Cropping systems based on diversity thus have a built-in protection.

Having destroyed nature's mechanisms for controlling pests through the destruction of diversity, the "miracle" seeds of the Green Revolution became mechanisms for breeding new pests and creating new diseases. The treadmill of breeding new varieties runs incessantly, as ecologically vulnerable varieties create new pests that create the need for breeding yet newer varieties.

The only miracle that seems to have been achieved with the breeding strategy of the Green Revolution is the creation of new pest and diseases, and with them the ever-increasing demand for pesticides. Yet the new costs of new pests and poisonous pesticides were never counted as part of the "miracle" of the new seeds that modern plant breeders had given the world in the name of increasing "food security."

The "miracle seeds" of the Green Revolution were meant to free the Indian farmer from constraints imposed by nature. Instead, large-scale monocultures of exotic varieties generated a new ecological vulnerability by reducing genetic diversity and destabilizing soil and water systems. The Green Revolution led to a shift from earlier rotations of cereals, oilseeds, and pulses to a paddy-wheat rotation with intensive inputs of irrigation and chemicals. The paddy-wheat rotation has created an ecological backlash with serious problems of waterlogging in canal-irrigated regions and groundwater mining in tubewell-irrigated regions. Further, the high-yielding varieties have led to large-scale micronutrient deficiencies in soils, particularly iron in paddy cultivation and manganese in wheat.

These problems were built into the ecology of the HYVs, even though they were not anticipated. The high water demands of these seeds necessitated high water inputs, and hence the hazards of desertification through waterlogging in some regions and desertification and aridization in others. The high nutrient demands caused micronutrient deficiencies, on the one hand, but were also unsustainable because increased applications of chemical fertilizers were needed to maintain yields, thus increasing costs without increasing returns. The demand of the HYV seeds for intensive and uniform inputs of water and chemicals also made large-scale monocultures an imperative, and monocultures being highly vulnerable to pests and diseases, a new cost was created for pesticide applications. The ecological instability inherent in HYV seeds was thus translated into economic nonviability. The miracle seeds were not such a miracle after all.

Sustainable agriculture is based on the recycling of soil nutrients. This involves returning to the soil part of the nutrients that come from the soil, either directly as organic fertilizer or indirectly through the manure from farm animals. Maintenance of the nutrient cycle, and through it the fertility of the soil, is based on this inviolable law of return, which is a timeless, essential element of sustainable agriculture.

The Green Revolution paradigm substituted the nutrient cycle with linear flows of purchased inputs of chemical fertilizers from factories and marketed outputs of agricultural commodities. Yet the fertility of soils cannot be reduced to N-P-K (nitrogen [N], phosphorus [P], and potassium [K]) in factories, and agriculture productivity necessarily includes returning to the soil part of the biological products that the soil yields. Technologies cannot substitute for nature and work outside nature's ecological processes without destroying the very basis of production. Nor can markets provide the only measure of "output" and "yields."

The Green Revolution created the perception that soil fertility is produced in chemical factories, and agricultural yields are measured only through marketed commodities. Nitrogen-fixing crops like pulses were therefore displaced. Millets, which have high yields from the perspective of returning organic matter to the soil, were rejected as "marginal" crops. Biological products not sold on the market but used as internal inputs for maintaining soil fertility were totally ignored in the cost-benefit equations of the Green Revolution miracle. They did not appear in the list of inputs because they were not purchased, and they did not appear as outputs because they were not sold.

Yet what is "unproductive" and "waste" in the commercial context of the Green Revolution is now emerging as productive in the ecological context and as the only route to sustainable agriculture. By treating essential organic inputs that maintain the integrity of nature as "waste," the Green Revolution strategy ensured that fertile and productive soils are actually laid waste. The "land-augmenting" technology has proved to be a land-degrading and land-destroying technology. With the greenhouse effect and global warming, a new dimension has been added to the ecologically destructive effect of chemical fertilizers. Nitrogen-based fertilizers release to the atmosphere nitrous oxide, which is one of the greenhouse gases causing global warming. Chemical farming has thus contributed to the erosion of food security through the pollution of land, water, and the atmosphere.[37]

Democratizing Knowledge

Modern silviculture as an exclusivist knowledge system that focuses exclusively on industrial wood production displaces local knowledge systems that view the forest in the perspective of food production, fodder production, and water production. The exclusive focus on industrial wood destroys the food-, fodder-, and water-production capacities of the forest. It disrupts links between forestry and agriculture, and in attempting to increase commercial/industrial wood, it creates a monoculture of tree species. The eucalyptus has become a symbol of this monoculture.

Modern agriculture focuses exclusively on agricultural commodity production. It displaces local knowledge systems that view agriculture as the production of diverse food crops with internal inputs, and replaces it with monocultures of introduced varieties needing external industrial inputs. The exclusive focus on external inputs and commercial outputs destroys diverse food crops such as pulses, oilseeds, and millets; disrupts the local ecological cycles; and in attempting to increase single-crop output, creates monocultures of crop varieties. The HYV becomes a symbol of this monoculture.

The crises of the dominant knowledge system has many facets.

1. Since dominant knowledge is deeply wedded to economism, it is unrelated to human needs. Ninety percent of such production of knowledge could be stopped without any risk of human deprivation. On the contrary, since a large part of such knowledge is a source of hazards and threats to human life (Bhopal, Chernobyl, Sandoz), its end would improve the possibilities of human well-being.
2. The political implications of the dominant knowledge system are inconsistent with equality and justice. It is disrupting of cohesion within local communities and polarizes society into those with access and those without it, both in respect to the knowledge systems and the power system.

3. Being inherently fragmenting and having built-in obsolescence, dominant knowledge creates an alienation of wisdom from knowledge and dispenses with the former.
4. It is inherently colonizing, inherently mystifying, shielding colonization by mystification.
5. It breaks away from concrete contexts, disqualifying as inadequate the local and concrete knowledge.
6. It closes access and participation to a plurality of actors.
7. It leaves out a plurality of paths to knowing nature and the universe. It is a monoculture of the mind.

Modern Western knowledge is a particular cultural system with a particular relationship to power. It has, however, been projected as above and beyond culture and politics. Its relationship with the project of economic development has been invisible; and therefore it has become a more effective legitimizer for the homogenization of the world and the erosion of its ecological and cultural richness. The tyranny and hierarchy privileges that are part of the development drive are also part of the globalizing knowledge in which the development paradigm is rooted and from which it derives its rationalization and legitimization. The power by which the dominant knowledge system has subjugated all others makes it exclusive and undemocratic.

Democratizing knowledge becomes a central precondition for human liberation because the contemporary knowledge system excludes the humane by its very structure. Such a process of democratization would involve a redefining of knowledge such that the local and diverse become legitimate as knowledge, viewed as indispensable because concreteness is the reality, and globalization and universalization are more mere abstractions that have violated the concrete and hence the real. Such a shift from the globalizing to the local knowledge is important to the project of human freedom because it frees knowledge from dependency on established regimes of thought, making it simultaneously more autonomous and more authentic. Democratization based on such an "insurrection of subjugated knowledge" is both a desirable and a necessary component of the larger processes of democratization because the earlier paradigm is in crisis and, in spite of its power to manipulate, is unable to protect both nature and human survival.

Notes

1. S. Harding, *The Science Question in Feminism* (Ithaca: Cornell University Press, 1986), 8.

2. T. Kuhn, *The Structure of Scientific Revolutions* (Chicago: University of Chicago Press, 1972).

3. R. Horton, "African Traditional Thought and Western Science," *Africa* 37, no. 2 (1967).

4. Harding, *The Science Question*, 30.

5. V. Shiva, *Ecology and the Politics of Survival* (Tokyo: United Nations University Press; New Delhi, London, Newbury Park, CA: Sage, 1991).

6. C. Caufield, *In the Rainforest* (London: Picador, 1986), 60.

7. E. Hong, *Natives of Sarawak* (Malaysia: Institut Masyarakat, 1987).

8. S. C. Chin, *The Sustainability of Shifting Cultivation* (Penang: World Rainforest Movement, 1989).

9. J. de Beer and M. McDermott, *The Economic Value of Non-timber Forest Products in Southeast Asia* (Amsterdam: Netherlands Committee for IUCN, 1989).

10. V. Shiva, *Staying Alive* (London: Zed Books, 1988), 59.

11. M. S. Randhawa, *A History of Agriculture in India* (New Delhi: Indian Council of Agricultural Research, 1989), 97.

12. K. K. Panday, *Fodder Trees and Tree Fodder in Nepal* (Berne: Swiss Development Cooperation, 1982).

13. S. P. Singh and A. Berry, *Forestry Land Evaluation at District Level* (Bangkok: FAO, 1985).

14. T. B. S. Mahat, *Forestry—Farming Linkages in the Mountains* (Kathmandu: ICIMOD, 1987).

15. World Rainforest Movement, *The Battle for Sarawak's Forests* (Penang: World Rainforest and SAM, 1990).

16. S. Schlich, *Systems of Silviculture* (1920).

17. R. S. Troup, *Silviculture Systems* (Oxford: Oxford University Press, 1916).

18. J. Bethel, "Sometimes the Word Is 'Weed,'" *Forest Management,* June 1984, 17–22.

19. V. Shiva, J. Bandyopadhyay, and H. C. Sharatchandra, *The Social, Ecological and Economic Impact of Social Forestry in Kolar* (Bangalore: IIM, 1981).

20. T. M. Quereshi, "The Concept of Fast Growth in Forestry and the Place of Indigenous Fast Growing Broad Leaved Species," in *Proceedings of the Eleventh Silvicultural Conference* (Dehra Dun: FRI, 1967).

21. A. N. Chaturvedi, *Eucalyptus for Farming* (Lucknow: U P Forest Department, 1983).

22. V. J. Patel, *Rational Approach towards Fuelwood Crisis in Rural India* (Surendrabag-Kardej: Jivarajbhai Patel Agroforestry Centre, 1984), 10.

23. R. K. Gupta, M. C. Aggarwal, and Hira Lal, "Correlation Studies of Phy-

tomass of Fodder Trees with Growth Parameters," *Soil Conservation Bulletin* (1984): 9.

24. R. V. Singh, *Fodder Trees of India* (New Delhi: Oxford University Press, 1982).

25. A. K. Yegna Iyengar, *Field Crops of India* (Bangalore: BAPPCO, 1944).

26. M. S. Swaminathan, *Science and the Conquest of Hunger Concept* (Delhi, 1983), 113.

27. C. H. Shah, ed., *Agricultural Development of India* (Delhi: Orient Longman, 1979), xxxii.

28. R. H. Richaria, paper presented at seminar Crisis in Modern Science, Penang, 1986.

29. Yegna Iyengar, *Field Crops of India,* 30.

30. Schlich, *Systems of Silviculture.*

31. Caufield, *In the Rainforest,* 177.

32. UNESCO, *Tropical Forest Ecosystems* (Paris: UNESCO, 1985).

33. Caufield, *In the Rainforest,* 178.

34. FAO, *Tropical Forest Management* (Rome: FAO, 1986).

35. L. Fahser, "The Ecological Orientation of the Forest Economy," lecture given at Faculty of Forestry Science, University of Freiburg, 1986.

36. V. Shiva and J. Bandyopadhyay, *Ecological Audit of Eucalyptus Cultivation* (Dehra Dun: Research Foundation, 1985).

37. V. Shiva, *The Violence of the Green Revolution* (Dehra Dun: Research Foundation of Science and Ecology, 1989).

6

Toward a New Agriculture Paradigm

Health per Acre

The Old Paradigm of Agriculture

The old paradigm of food and agriculture is clearly broken. As the report of the International Assessment of Agricultural Knowledge, Science and Technology for Development (IAASTD) carried out by four hundred scientists over six years for the United Nations has noted, "Business as usual is no longer an option."

The old paradigm of agriculture has its roots in war. An industry that had grown by making explosives and chemicals for the war remodeled itself as the agrochemical industry when the wars ended. Explosive factories started to make synthetic fertilizers; war chemicals started to be used as pesticides and herbicides. The Bhopal tragedy in 1984, when a gas leak from a pesticide plant killed three thousand people, thirty thousand since then, is a stark reminder that pesticides kill. Pesticides in agriculture continue to kill farm workers. And as the Navdanya report *Poisons in Our Food* shows, there is a link between disease epidemics like cancer and the use of pesticides in agriculture. A daily "cancer train" leaves Punjab, the land of the Green Revolution in India, with cancer victims.

The chemical push changed the paradigm of agriculture. Instead of working with ecological processes and taking the well-being and health of the entire agroecosystem with its diverse species into account, agriculture was reduced to an external input system adapted to chemicals. Instead of recognizing that farmers have been breeders over millennia, giving us the rich agrobiodiversity that is the basis of food security, breeding was reduced to breeding uniform industrial varieties to respond to chemicals. Instead of small farms producing diversity, agriculture became focused on large monoculture farms producing monocultures of a handful of commodities. Correspondingly, the human diet shifted from the use of eighty-five hundred plant species to about eight globally traded commodities.

The scientific paradigm was also transformed. Instead of a holistic approach, agriculture became compartmentalized into fragmented disciplines based on a reductionist, mechanistic paradigm.

Just as GDP fails to measure the real economy, the health of nature and society, the category of "yield," designed to measure the productivity of agriculture, fails to measure real costs and real outputs of farming systems. It leaves out input costs, which if internalized would make industrial agriculture based on the old paradigm a negative economy, using ten times more inputs than it produces as a commodity. Further, it leaves out the outputs lost, as chemical-driven agriculture imposes monocultures and destroys diversity. In India, the Green Revolution drove out pulses and oilseeds and greens from the farming system. Rice and wheat production increased, but pulses and oilseeds disappeared. The studies of Navdanya/Research Foundation for Science, Technology and Ecology show that the increase of rice and wheat is explained by more acreage of rice and wheat, and more irrigation water made available. It is not a result of so-called miracle seeds and agrochemicals. Leaving out the external inputs and the biodiverse outputs thus creates a pseudo-productivity, making industrial agriculture appear the only solution to hunger, when in reality it is at the root of hunger and malnutrition by destroying sources of food in biodiversity. And even the "yield" of a monoculture is an unreliable measure. As the UN observed, the so called high-yielding varieties of the Green Revolution should in fact be called high-response varieties since they are bred for responding to chemicals and are not high yielding in and of themselves.

The narrow measure of "yield" propelled agriculture into deepening monocultures, displacing diversity and eroding natural and social wealth. The social and ecological impacts of this broken model have pushed the planet and society into deep crisis. Industrial monoculture agriculture has caused the extinction of more than 75 percent of agrobiodiversity. Seventy-five percent of bees have been killed by toxic pesticides. Einstein had cautioned "when the last bee disappears, humans will disappear." Seventy-five percent of the water on the planet is being depleted and polluted for intensive irrigation for chemical intensive industrial agriculture. The nitrate in water from industrial farms is creating "dead zones" in the oceans.[1] Seventy-five percent of land and soil degradation is caused by chemical industrial farming. Forty percent of all greenhouse gas emissions responsible for climate change come from a fossil fuel, chemical-intensive industrial globalized system of agriculture. The fossil fuels used to make fertilizers, run farm machinery, and move food thousands of miles contribute to carbon dioxide emissions. Chemical nitrogen fertilizers emit nitrogen oxide, which is 300 percent more destabilizing for the climate than carbon dioxide, and factory farming is a major source of methane.[2]

Although this ecological destruction of nature's economy is justified in terms of "feeding people," the problem of hunger has grown: 1 billion people are permanently hungry. Another 2 billion suffer from food-related diseases like obesity, which are being increasingly related to micronutrient deficiencies.

When the focus is the production of commodities for trade instead of food for nourishment, hunger and malnutrition are the outcome. Only 10 percent of the corn and soy grown is used as food. The rest goes for animal feed and biofuel. Commodities do not feed people; food does. Seventy percent of food comes from small family farms, not industrial commodity-producing farms. To expand industrial farming and genetically modified organisms (GMOs) in the name of feeding the hungry is a recipe for increasing the food and malnutrition crisis.

A high-cost external input system is artificially kept afloat with $400 billion in subsidies. That is more than $1 billion a day.

The "cheap" commodities have a very high cost financially, ecologically, and socially. Industrial, chemical agriculture displaces productive rural families. It creates debt; debt and mortgages are the main reasons for the disappearance of the family farm. In extreme cases in India, as in the cotton belt, debt created by the purchase of high-cost seed and chemical inputs has pushed more than 128,000 farmers to suicide in a little over a decade. Getting out of this suicide economy has become urgent for the well-being of farmers, eaters, and all life on earth.

The Emerging Paradigm of Agriculture

A scientifically and ecologically robust paradigm of agriculture is emerging in the form of agroecology and organic farming as an alternative to the broken paradigm of industrial agriculture. At the ecological level, agroecology and organic farming rejuvenate nature's economy, on which sustainable food security depends—soil, biodiversity, and water.

Chemical agriculture treats soil as inert and an empty container for chemical fertilizers. The new paradigm recognizes the soil as living, host to billions of soil organisms that create soil fertility. Their well-being is vital to human well-being. Chemical agriculture destroys biodiversity. Ecological agriculture conserves and rejuvenates biodiversity, and through biodiversity intensification, it increases the food and nutrition output, or health per acre. Chemical agriculture depletes and pollutes water. Organic farming conserves water by increasing the water-holding capacity of soils through recycling organic matter. The soil becomes like a sponge, which can absorb more water, thus reducing water use, but also contributing to resilience to climate change. Biodiversity and soils rich in organic matter are the best strategy for climate resilience and climate adaptation.

While rejuvenating natural capital, ecological agriculture also rejuvenates social capital and increases human well-being and happiness. While reducing the ecological footprint, organic agriculture increases output when measured through multifunctional benefits instead of the reductionist category of "yield." As Navdanya's research on biodiverse organic systems has shown, ecological systems produce higher biodiverse outputs and higher incomes for rural families. Our report *Health per Acre* shows that when measured in terms of nutrition per acre, ecological systems produce more food. We can double food production ecologically. The false argument that GMOs are needed to increase food production is a desperate attempt to extend the life of a failing paradigm. The new paradigm of agriculture creates living economies, living democracies, and living cultures that are the foundation for earth democracy and increase the well-being of society and all life-forms.

India faces a dual crisis related to food and agriculture. We have already touched on the agrarian crisis, tragically highlighted by farmers' suicides, driven by debt that is largely caused by high-cost chemical inputs. The other aspect is the malnutrition and hunger crisis. Every fourth Indian is hungry.[3] Every third woman is severely malnourished. Every second child is "wasted." This is not "Shining India" but "Starving India." The agrarian crisis and the food and nutrition crisis are really connected.

Taking note of the hunger and malnutrition crisis, the government is trying to put together a Food Security Act. However, there are two serious limitations to the proposed act. First, it leaves out nutrition. Without nutrition there can be no right to food or health. Malnutrition is leading to a public health crisis—of hunger on the one hand, and obesity, diabetes, and other health problems on the other. Second, it leaves out agriculture, food producers, and food-production systems. Without agriculture and nutrition, there can be no food security.

Both aspects of the food crisis are related to the fact that food production has become chemical intensive and is focused on "yield per acre." However, yield per acre ignores the loss of nutrition that is leading to the malnutrition crisis. It also ignores the increase in costs of chemical inputs, which traps farmers in debt and leads to suicides. "Yield per acre" measures one crop grown in a monoculture. This ignores the lost nutrition in the displaced biodiversity. Thus the Green Revolution led to an increase of rice and wheat with chemical-intensive, capital-intensive, and water-intensive inputs, but it displaced pulses, oilseeds, millets, greens, vegetables, and fruits from the field and from the diet.

Navdanya's *Health per Acre* shows that a shift to biodiverse organic farming and ecological intensification increases output of nutrition while reducing input costs. When agriculture output is measured in terms of "health per acre" and "nutrition per acre" instead of "yield per acre," biodiverse ecological systems have a much higher output. This should be the strategy for protecting the livelihoods of farmers as well the right to food and the right to health of all our people.

The paradigm shift we propose is a shift from monocultures to diversity; from chemical-intensive agriculture to ecologically intensive, biodiversity-intensive agriculture; from external inputs to internal inputs; from capital-intensive production to low-cost or zero-cost production; from yield per acre to health and nutrition per acre; from food as a commodity to food as nourishment and nutrition. This shift addresses the multiple crises related to food systems. It shows how we can protect the environment while protecting our farmers and our health. And we can do this while lowering costs of food production and distribution. By maximizing health per acre, we can ensure that every child, woman, and man in India has access to healthy, nutritious, safe, and good food.

Food, nutrition, health, prosperity, future, and growth and their opposites—hunger, disease, poverty, hopelessness, and the nation's downfall—are much-debated topics that, intuitively, are not only correlated but also have a causal connection. Agriculture, a time-tested profession, one of the oldest in the world, is no longer an economically viable endeavor for most. However, the question to be answered is whether our nation is committing suicide as well as our farmers. The primary objective of a nation's agriculture is to promote health and feed the people, propagating a diet that provides all the necessary nutrients. However, profit maximization has been promoted as the objective of agriculture. Tragically, the more profit-oriented agriculture becomes, the higher the farmers' indebtedness and farmers' suicides, and the deeper the food and nutrition crisis. The irony is that, despite all the claims, maximization of profit for farmers is still far away from realization,[4] but the nation has been paying the enormous cost.

Most proponents of conventional agriculture claim that pesticides, one of the many chemicals used in agriculture, have insignificant implications for human health. Nevertheless, millions of tons of pesticides pumped into the environment every year in the name of high-yield agriculture somehow manage to reach the human body as well as animals and water supplies. Quantifiable levels of a number of pesticides have been detected in human milk, which puts breast-feeding infants at probable risk.[5] The alarming level of chemicals in the honey sold in Indian markets triggered much discussion recently. Science and technology were established to benefit human beings, but in current agricultural practice, they are benefiting human greed. As a major contributor to global warming, the conventional form of agriculture has negative health impacts as well. We shall limit our discussion in this chapter to the effects of conventional agriculture on the health of individuals and the population as a whole. This report compares the nutritional and health aspects of food grown organically and food grown conventionally. The scope of the work ranges from nutrition produced per acre of farmland by the two systems of agriculture to disease trends observed in the population and how such trends may be related to the food we consume. Conventional agriculture measures "yield" per acre while externalizing costs of chemical inputs and the environmental and health costs of chemicals. "Yield" measures monoculture outputs, while what we need to assess is diverse outputs of a farming system. Yield also fails to tell us about the nutrition of food. With a focus on health and nutrition, we measure health per acre instead of yield per acre.

Nutrition may be defined as the science of food and its relationship to health.[6] It is primarily related to the role played by nutrients in body growth, development, and maintenance. Good nutrition means "maintaining a nutritional status that enables us to grow well and enjoy good health." Nutrients are organic and inorganic complexes contained in food. Each nutrient has a specific function in the body. Nutrients may be classified as below:

1. Macronutrients: they form the main bulk of food. These are protein, carbohydrates, and fat.
2. Micronutrients: they are required in small amounts. These are vitamins and minerals.

There are several bioactive compounds in plant food, and several health benefits are attributed to the presence of such compounds in diet. Studies have shown that individuals with increased consumption of fruits and vegetables showed a lower incidence of chronic noncommunicable diseases such as cancer, cardiovascular diseases, diabetes, and age-related decline in cognition.[7] Scientists agree upon the health benefits of the consumption of fruits and vegetables. The American Heart Association and the American Cancer Society recommend a generous daily intake of fruits and vegetables.

Earlier it was thought that the health benefits of fruits and vegetables could be due to the antioxidant effects of various micronutrients present in high quantity in them. This highlighted the need for more research to isolate such protective compounds in plant food for therapeutic purposes. Scientists studied the incidence of different chronic diseases in individuals who consumed vitamin, mineral, and antioxidant supplements. Incidentally, these individuals were no healthier than the normal population in terms of incidence of various cancers, heart diseases, and other chronic diseases. Researchers were compelled to think outside the box. There was something extra in plant food that was unknown. Finally, such compounds as phytochemicals, phenols, flavonoids, and so on in plants were recognized as health-promoting chemicals. Studies have shown the link between these bioactive compounds and prevention of chronic noncommunicable diseases.[8] These compounds contribute significantly to the total antioxidant activity of fruits and vegetables. These compounds deliver an electron to reactive oxygen species (ROS, which are produced in the body as a result of stress, smoking, disease, and so on) and render them ineffective. ROS are highly reactive and damage cellular macromolecules (protein, membrane, DNA, RNA, and the like). ROS are thought to cause cancers, cardiovascular diseases, diabetes, and other chronic diseases in the long run.

Case Study 1

Under monocropping of paddy, a yield of twelve quarts per acre was observed, whereas under mixed cropping a production of three quarts of mandua (ragi), two quarts of jhangora (sanwa millet), four quarts of gahat (horse gram), and five quarts of bhatt (black bean or rajmah) was realized.

Organic mixed farming produced 276 percent more protein per acre of farmland than that produced by conventional monocropping. Organic mixed cropping produced 10,483 percent more carotene, 188 percent more thiamine, and 83 percent more riboflavin per acre of farmland than produced by conventional monocropping. Organic mixed cropping produced generous amounts of vitamin B, folic acid, and vitamin C, which conventional monocropping did not produce. However, conventional monocropping produced 39 percent more niacin per acre of farmland than that produced by organic mixed farming. The increase in the production of niacin and choline is attributed to the fact that paddy is a rich source of these vitamins and twelve quarts of paddy were produced.

The total amount of major minerals produced per acre of farmland in organic mixed cropping was 16,527.8 grams. The total amount of major minerals produced per acre of farmland in conventional mono-cropping was 4,322 grams. Organic mixed cropping produced 282 percent more major minerals per acre of farmland than produced by conventional monocropping. Moreover, organic mixed cropping produced 163 percent more iron per acre of farmland than conventional monocropping.

The total amount of trace minerals produced per acre of farmland in organic mixed cropping was 1,299,572 milligrams. The total amount of trace minerals produced per acre of farmland in conventional monocropping was 33,924 milligrams. Organic mixed cropping produced 3,731 percent more trace minerals than conventional monocropping.

Case Study 2

Organic mixed cropping produced 26 percent more protein per acre of farmland than conventional monocropping. Organic mixed cropping produced 3,000 percent more carotene and 88 percent more thiamine than conventional monocropping. Moreover, organic mixed cropping produced folic acid, vitamin B, and vitamin C, which conventional monocropping did not produce. However, monocropping produced more niacin and choline because paddy is a rich source of these vitamins.

The total amount of minerals produced per acre of farmland in organic mixed cropping was 12,696 grams. The total amount of minerals produced per acre of farmland in conventional monocropping was 4,322 grams.

Organic mixed cropping produced 194 percent more minerals than conventional monocropping per acre of farmland. Moreover, organic mixed cropping produced 27 percent more iron. The total amount of trace minerals produced per acre of farmland in organic mixed cropping was 15,63,918 milligrams. The total amount of trace minerals produced per acre of farmland in conventional monocropping was 33,924 milligrams. Organic mixed cropping produced 4,510 percent more trace minerals than conventional monocropping per acre of farmland.

Researchers and doctors globally have reached a collective consensus that one should derive one's nutrition from diverse sources. How will our meal plate, or *thali*, be diverse if our farms aren't? There is a concept in finance that emphasizes diversification of a portfolio to reduce risk; this seems an equally valuable concept for agriculture, health, and nutrition. According to Rui Hai Liu from the Department of Food Science, Cornell University, Ithaca, New York, "We believe that a recommendation that consumers eat 5 to 10 servings of a wide variety of fruits and vegetables daily is an appropriate strategy for significantly reducing the risk of chronic diseases and to meet their nutrient requirements for optimum health."[9] How can we expect to consume such a wide variety of foods if we do not grow such a wide variety?

Our per capita nutrition, or average nutrition per person per day, has declined significantly from 1975 to 1999. That period is also significant from the Green Revolution point of view—in 1975 the effects of the Green Revolution and conventional farming were negligible, whereas by 1999 conventional farming practices had gripped our society substantially. One probable reason for such a change in average nutritional consumption is the population explosion. However, to blame everything on the rise in population would be shortsighted and superficial. Further extensive research is required to prove a definite correlation.

Another interesting fact that came out was that an acre of farmland under conventional agriculture produced low amounts of most nutrients. However, such farmland produced a few odd nutrients excessively. This has probably affected our national health; on the one hand we are struggling to treat and eradicate deficiency diseases like protein energy malnutrition, night blindness, anemia, and so on, and on the other hand the nation is distressed by the debilitating effects of excessive nutrition, such as obesity, hypervitaminosis, cardiovascular diseases, diabetes, and the like. However, in order to prove a definite correlation, further extensive research is called for.

Diversification is not important just from the "amount of nutrient produced per acre" point of view. Research has suggested that traditional foods and different varieties of fruits and vegetables contain several bioactive compounds that prevent cancer, diabetes, cardiovascular diseases, and other degenerative diseases. All such compounds have not been identified to date, the role of such bioactive compounds in preventing these degenerative diseases has not yet been pinpointed, and an ideal blend of nutrients for human consumption has not been determined. We are almost there, but not quite. As a result, medical practitioners prescribe a diet derived from varied sources.[10]

In order to provide a more comprehensive picture, we took the average (arithmetic mean) of nutrients produced per acre of farmland from the case studies above. The sample mean of our report should be a fairly good estimator of the population mean. The population in our case is the total arable land in India. Hence, the average production of nutrients per acre of farmland is a reasonably fair point estimator of the average production per acre of farmland on a national scale. Moreover, we have collected data from different states ranging from an arid state, Rajasthan, to an organic state, Uttaranchal. As a result the margin of error should be fairly low. The purpose of the statistics is to allow the reader to glimpse the actual effect of the two forms of agriculture on a national level. The questions are how to maximize nutrient production, how to minimize environmental risk, and how to ensure a sustainable alternative to solve the national and global food crisis.

If we switch an acre of farmland from conventional monocropping to organic mixed cropping, we shall be able to produce 124 more kilograms of protein. The quality of mixed-cropping protein is better than that of monocropping protein. The organic mixed-cropping protein is complete because it provides all the essential amino acids—it is comparable to animal protein. Vegetarian protein (except soy) may be an inadequate source of all essential amino acids individually. However, when vegetarian proteins are mixed, they become an adequate source of all essential amino acids. For example, the protein in roti or dal, individually, is incomplete because it does not contain all the essential amino acids, but when roti and dal are consumed together, they become a complete source of all essential amino acids.[11] Hence, the protein produced in an acre of farmland under organic mixed cropping is more complete than protein produced in an acre under conventional monocropping.

On average, organic mixed cropping produces 124 more kilograms of protein than conventional monocropping per acre of farmland, enough to fulfill the protein requirement of two thousand adults for a day. According to the Central Water Commission, government of India, the total cultivable land (as of 2003–2004) in India is 183 million hectares, which is equal to approximately 452,202,848 acres. If all of this land is used for organic mixed cropping instead of conventional monocropping, the country can produce 56,073,153 more metric tons of protein. This is enough to fulfill the protein requirement of 2.5 billion adults for the entire year. A fact worthy of notice is that we have counted here only the *difference* of 124 kilograms of protein per acre between organic mixed cropping and conventional monocropping. If we consider the *entire* amount of protein produced in the country through organic mixed cropping, by projecting our sample average to the total cultivable land, we would produce enough protein to fulfill the protein requirement of approximately 5 billion adults for a whole year. This is enough protein to feed our entire population and to eradicate protein energy malnutrition from the planet.

If an acre of farmland is diverted from conventional monocropping to organic mixed cropping, we shall produce additional food containing 12,02,795 kilocalories of extra energy. This is enough to supply 2,500 kilocalories of energy to 481 adults for a day. If we project this figure to 183 million hectares of total cultivable land in India, we shall produce additional calories in food sufficient to fulfill the energy requirement of 600 million adults for the whole year. We would again like to emphasize that we considered only the extra calories produced by switching from conventional to organic. If we consider the sample average amount of calories produced per acre through organic mixed cropping, then on a national scale, we shall produce enough calories to supply 2,500 kilocalories a day to 2.4 billion adults for one year. If we switch from conventional to organic, we can ensure that no individual is hungry in our country. In fact, if only India switches from conventional agriculture to organic agriculture, we can resolve the global hunger problem because it is just the bottom billion of the world population that is hungry.

If an acre of farmland is used for organic mixed cropping rather than conventional monocropping, we shall produce 2,174 milligrams of carotene more than that produced otherwise. This is enough carotene to fulfill the vitamin A requirement of approximately 900 adults for a day. On a national scale, we would produce 982,670 more metric tons of carotene organically than produced conventionally. In other words, we would produce 164,106 more metric tons of retinol equivalent (RE) (1 unit of B-carotene = 0.167 unit of RE) than produced conventionally.[12] That amount is sufficient to satisfy the daily vitamin A requirement of 750 million adults for one year. It is sufficient to completely reverse 1.3 billion early cases of xerophthalmia, assuming that all this retinol equivalent in food can be isolated and administered to xerophthalmia patients. The term *xerophthalmia* (dry eye) comprises all the ocular manifestations of vitamin A deficiency, ranging from night blindness to keratomalacia. Vitamin A deficiency first causes night blindness and then progresses to corneal ulcers—a serious condition that may leave residual corneal scar, affecting vision. Keratomalacia, or liquifaction of the cornea, is a major cause of blindness in India—the cornea becomes soft and may burst open. This is an example of the kind of significant impact that switching to organic on a national scale could have on the health of our population. If we use the sample average amount of carotene produced per acre of farmland by organic mixed cropping to calculate the total amount of carotene produced nationally, we can produce enough to fulfill the daily vitamin A requirement of 1.5 billion adults for one year.

Similarly, the extra amount of thiamine produced per acre by switching from conventional to organic is enough to supply thiamine to approximately 2,100 adults for a day. On a national scale, the extra amount of thiamine produced by switching from conventional to organic would be sufficient to fulfill the daily thiamine requirement of 2.6 billion adults for one year. If we consider all the thiamine that can be produced organically in the country, then the thiamine produced would be sufficient for approximately 5 billion adults for a year. Minor degrees of thiamine deficiency is endemic in certain sections of the country.[13] With organic farming on a national scale, we can uproot and eradicate all forms of thiamine deficiency from our population.

Organic mixed cropping in an acre of farmland produces extra riboflavin, compared to conventional monocropping, that can fulfill the recommended riboflavin allowance of 1,000 adults for a day. On a national scale, we could supply adequate amounts of riboflavin to 1.2 billion extra adults for a year. Riboflavin deficiency is widespread in India, particularly where rice is the staple.[14] The fact is that we are not currently producing enough riboflavin; organic mixed cropping seems to be a promising solution to resolve the riboflavin crisis.

Folic acid deficiency can occur rapidly in pregnant and lactating mothers and growing children because body stores of folate are not large—about five to ten milligrams. An acre of farmland through organic mixed cropping can produce extra folic acid that can nourish approximately 1,375 pregnant mothers for a day. On a national scale, the extra amount of folate produced through organic mixed cropping compared to its conventional counterpart is sufficient to supply folic acid to 1.7 billion pregnant woman, who require four times as much folic acid as a normal adult, for one year.

Our sample shows that vitamin C produced by conventional mono-cropping was more than that produced by organic mixed cropping. Nevertheless, there are a few points that need to be highlighted. Although the mean production of vitamin C of our sample favors conventional mono-cropping, the median value is zero in conventional monocropping compared to organic mixed cropping, which has a median value of 4,470 milligrams. We extrapolate from this that a farmer in Rajasthan or Sikkim, practicing conventional monocropping, would suffer from vitamin C deficiency, whereas a farmer in Uttaranchal who produced excess vitamin C would expel the excess in his urine—we assumed that the farmers consumed only the food that they grew.

According to Virginia Worthington's research, organically grown food has 27 percent more vitamin C, on an average, than conventionally grown food.[15] If we include the difference of 27 percent in our sample mean, the difference decreases drastically.

Iron is of great importance to human health. The adult human body contains about three to four grams of iron, of which 60–70 percent is present in blood. Iron is required for many functions in the body, such as hemoglobin formation, brain development and function, regulation of body temperature, muscle activity, and catecholamine metabolism. The central function of iron is oxygen transport and cell respiration. The bioavailability of nonhaem iron (mostly vegetarian) is poor owing to the presence of phytates, oxalates, carbonates, phosphates, and dietary fiber. The Indian diet, which is predominantly vegetarian, contains large amounts of such inhibitors—phytates in bran, phosphates in egg yolk, tannin in tea, and oxalates in vegetables. Deficiency of iron in diet leading to iron deficiency anemia or nutritional anemia is a major public health problem in India.

When an acre of farmland is used for organic mixed cropping in place of conventional monocropping, thirty-nine grams of extra iron is produced. This amount is sufficient to nourish 16,250 lactating mothers with iron for a day. On a national scale, the extra amount of iron produced organically would be sufficient to meet the requirement of 20 billion hypothetical lactating mothers. To reach this conclusion, we assumed that all of the iron consumed would be absorbed.

Organic mixed cropping, on average, produces 106 percent more copper, 61 percent more manganese, 243 percent more molybednum, 64 percent more zinc, and 120 percent more chromium than conventional monocropping. Collectively, organic mixed cropping produces 72 percent more of these trace minerals than conventional monocropping does. Micronutrient deficiency is increasingly being observed in soil and in humans.

Biodiverse Ecological Systems Produce More Food

Just as the food crisis is a consequence of a food system designed for profits, greed, and control, we can redesign the food system for sustainability and food justice. And this redesigning is precisely what we are doing at Navdanya. Over twenty years of research and practice, we are finding that biodiverse ecological production systems are the solution to hunger and malnutrition, to the agrarian crisis and farmers' suicides, to the erosion of soil, water, and biodiversity, and to the climate crisis.

The Green Revolution and genetic engineering have been offered as "intensive" farming, creating a false impression that they produce more food per acre. However, industrial agriculture is *chemically intensive, fossil fuel intensive,* and *capital intensive.* The first two qualities produce more toxics and greenhouse gases and the third more debt.

To produce more food and nutrition, we need to design production systems that are *biodiversity intensive* and *ecologically intensive.* Biodiversity-intensive systems produce more food, nutrition, and health per acre than industrial chemical monocultures. And by saving on costs of external inputs, they create more wealth per acre for farmers. When measured in terms of contribution to nutrition, health, and rural incomes, industrial systems have very low productivity. Not only is the measure of productivity of industrial agriculture partial because all inputs, including resource and energy inputs, are not taken into account, but it is also partial because not all outputs are taken into account. Only the production of monoculture commodities is counted.

Green Revolution systems have high "yield" but low output. And it is output that feeds the soil and people, not the yield of globally traded commodities that are used for biofuel or animal feed. Ecological agriculture is based on mixed and rotational cropping, and the production of a diversity of crops. Navdanya's work on biodiverse farming has shown that the more biodiversity on the farm, the higher the output.[16]

Perhaps one of the most fallacious claims propagated by Green Revolution proponents is the assertion that HYVs have reduced the acreage necessary to grow these crops, therefore preserving millions of hectares of biodiversity. Perpetuating this myth, Dennis Avery, a promoter of chemical farming, has recently written, "Is the Green Movement finally ready to face the global need to triple crop yields and drop its dedication to land selfish organic farming? The planet's biodiversity is at stake." India's experience tells us that instead of more land being released for conservation, by destroying diversity and multiple uses of land, the industrial system actually increases pressure on the land since each acre of a monoculture provides a single output and the displaced outputs have to be grown on additional acres. And globally, the chemical-intensive, land-extensive system has had to spread to the Amazon rainforest. This is not land-saving, biodiversity-conserving agriculture—it is land-destroying, biodiversity-destroying agriculture.

The polycultures of ecological agricultural systems have evolved because more output can be harvested from a given area planted with diverse crops than from an equivalent area consisting of separate patches of monocultures. For example, in plantings of sorghum and pigeon pea mixtures, one hectare will produce the same yields as 0.94 hectares of sorghum monocultures and 0.68 hectares of pigeon pea monoculture. Thus one hectare of polyculture produces what 1.62 hectares of monoculture can produce. This is called the land-equivalent ratio (LER).

Increased land-use efficiency and higher LER have been reported for polycultures of millet/groundnut (1.26); maize/bean (1.38); millet/sorghum (1.53); maize/pigeon pea (1.85); maize/cocoyan/sweet potato (2.08); cassava/maize/groundnut (>2.51). The monocultures of the Green Revolution thus actually reduced the food yields per acre previously achieved through mixtures of diverse crops. This shows the falsity of the argument often made that chemically intensive agriculture and genetic engineering will save biodiversity by releasing land from food production. In fact, since monocultures require more land, biodiversity is destroyed twice over—once on the farm, and then on the additional acreage required to produce the outputs a monoculture has displaced. Further, since chemicals kill diverse species, chemical agriculture can hardly be promoted as conserving biodiversity.

Not only is the productivity measure distorted by ignoring resource inputs and focusing only on labor, but it is also distorted by looking only at a single and partial output rather than the total output. A myth promoted by the one-dimensional monoculture paradigm is that biodiversity reduces yields and productivity, and monocultures increase yields and productivity. However, since yields and productivity are theoretically constructed terms, they change according to the context. "Yield" usually refers to production per unit area of a single crop. Planting only one crop in the entire field as a monoculture will of course increase its yield. Planting multiple crops in a mixture will have low yields of individual crops but will have high total output of food.

The Mayan peasants in the Chiapas are characterized as unproductive because they produce only two tons of corn per acre. However, the overall food output is twenty tons per acre. In the terraced fields of the high Himalaya, women peasants grow jhangora (barnyard millet), marsha (amaranth), tur (pigeon pea), urad (black gram), gahat (horse gram), soybean (glysine max), bhat (glysine soy), rayans (rice bean), swanta (cowpea), and koda (finger millet) in mixtures and rotations. The total output, even in bad years, is six times more than industrially farmed rice monocultures.

Not only do biodiverse-intensive and ecologically intensive systems produce more food per acre, but they also produce much higher nutrition per acre. For example, a mixed organic farm in the Himalaya produces 9,000 kilograms of maize, radish, mustard greens, and peas. A chemically farmed maize monoculture yields 5,000 kilograms. This is 1,000 kilograms more maize than in the biodiverse system, but 4,000 kilograms less food. In terms of nutrition per acre, the biodiverse farming system is much more productive than the chemical monocultures. It provides 305 grams of calcium and 29.3 grams of iron compared to the monoculture.

Similarly, a biodiverse-intensive system with mandua (finger millet), jhangora (barnyard millet), gahat (horse gram), and bhatt (indigenous soy) gives 1,400 kilograms of food per acre compared to a chemical rice monoculture, which yields 1,200 kilograms. In terms of nutrition, the former gives 338 kilograms of protein compared to 90 kilograms in the monoculture. The biodiverse-intensive system gives 2,540 milligrams of carotene compared to 24 milligrams in the monoculture, and 554 milligrams of folic acid compared to 0 in the rice monoculture. Calcium is 3,420 grams compared to 120 grams. Iron is 100.8, compared to 38.4; phosphorous is 6,103, compared to 2,280; magnesium is 2,389, compared to 1,884; and potassium is 4,272, compared to 0.

A *baranaja* (twelve-crop) system produces 2,680 kilograms of food per acre, compared to 2,186 of a maize monoculture. In terms of protein, the production is 4,214 versus 242 kilograms; carbohydrates, 1,622.94 versus 1,447.14 kilograms; fat, 131.8 versus 78.7 kilograms; energy, 9,359,470 versus 7,476,120 kilocalories. In terms of vitamins, baranaja produces 1,360.9 versus 1,967 milligrams beta carotene in the case of maize monoculture; in folic acid, 2,206.3 to 437 milligrams. Minerals are: calcium, 5,052 versus 218 grams; iron, 143.9 versus 50.3 grams; phosphorous, 9,505 versus 7,607 grams; magnesium, 3,604 versus 3,038 grams; potassium, 11,186 versus 6,252 grams.[17]

Since providing nutrition and nourishment are the main aims of agriculture, in food production, nutrition per acre is a more accurate measure of productivity than the yield of commodities in a monoculture. Also, the higher nutrition in biodiverse-intensive farms further intensifies the ecological processes. The nutrients produced by plants become food for humans and food for soil organisms, which in turn feed the plants that feed the humans and the soils. The perennial nutrient cycle continues to be sustained and can even be intensified through biodiversity intensification and ecological intensification.

A model of nutrients for soils based on heavy inputs of nonrenewable N-P-K impoverishes the soil, our diets, and our health. In any case, industrial sources of nonrenewable N-P-K are running out. Ecological nutrients are renewable; they will last forever, and we can actually increase their availability by increasing the biodiversity of soil organisms and plants.

The main argument used for the industrialization of food and corporatization of agriculture is the low productivity of the small farmer. Surely these families on their little plots of land are incapable of meeting the world's need for food! Industrial agriculture claims that it increases yields, hence creating the image that more food is produced per unit acre by industrial means than by the traditional practices of smallholders. However, sustainable, diversified small-farm systems are actually more productive.

Industrial agriculture productivity is high only in the restricted context of a "part of a part" of the system, whether it be the forest or of the farm. For example, "high-yield" plantations pick one tree species among thousands, for yields of one part of the tree (for example, wood pulp), whereas traditional forestry practices use many parts of many forest species. "High-yield" Green Revolution cropping patterns select one crop among hundreds, such as wheat, for the use of just one part, the grain. These high partial yields do not translate into high total yields, because everything else in the farm system goes to waste. Usually the yield of a single crop like wheat or maize is singled out and compared to yields of new varieties. This calculation is biased to make the new varieties appear "high yielding" even when, at the systems level, they may not be. Traditional farming systems are based on mixed and rotational cropping systems of cereals, pulses, and oilseeds, with different varieties of each crop, while the Green Revolution package is based on genetically uniform monocultures. No realistic assessments are ever made of the yield of the diverse crop outputs in the mixed and rotational systems.

Productivity is quite different, however, when it is measured in the context of diversity. Biodiversity-based measures of productivity show that small farmers can feed the world. Their multiple yields result in truly high productivity, composed as they are of the multiple yields of diverse species used for diverse purposes. Thus, productivity is not lower on smaller units of land: on the contrary, it is higher. In Brazil, the productivity of a farm of up to ten hectares was $85 a hectare, while the productivity of a five-hundred-hectare farm was $2 per hectare. In India, a farm of up to five acres had a productivity of Rs. 735 per acre, while a thirty-five-acre farm had a productivity of Rs. 346 per acre.

Diversity produces more than monocultures. But monocultures are profitable to industry both for markets and political control. The shift from high-productivity diversity to low-productivity monocultures is possible because the resources destroyed are taken from the poor, while the higher commodity production brings benefits to those with economic power. The polluter does not pay in industrial agriculture either of the chemical era or the biotechnology era. Ironically, while the poor go hungry, it is the hunger of the poor that is used to justify the very agricultural strategies that deepen their hunger.

Diversity has been destroyed in agriculture on the assumption that it is associated with low productivity. This is, however, a false assumption both at the level of individual crops and at the level of farming systems. Diverse native varieties are often as high yielding or more high yielding than industrially bred varieties. In addition, diversity in farming systems has a higher output at the total systems level than one-dimensional monocultures. Comparative yields of native and Green Revolution varieties in farmers' fields have been assessed by Navdanya, a seed conservation and agroecology movement. Green Revolution varieties are not higher yielding under the conditions of low capital availability and fragile ecosystems. Farmers' varieties are not intrinsically low yielding, and Green Revolution varieties or industrial varieties are not intrinsically high yielding.

The measurement of yields and productivity in the Green Revolution as well as in the genetic engineering paradigm is divorced from seeing how the processes of increasing single-species, single-function output affect the processes that sustain conditions of agricultural production, both by reducing species and functional diversity of farming systems and by replacing internal inputs provided by biodiversity with hazardous agrochemicals. While these reductionist categories of yield and productivity allow a higher measurement of harvestable yields of single commodities, they exclude the measurement of the ecological destruction that affects future yields and the destruction of diverse outputs from biodiversity-rich systems.

Productivity in ecological farming practices is high if it is remembered that these practices are based on internal inputs, with very little external input required. While the Green Revolution has been projected as having increased productivity in the absolute sense, when resource utilization is taken into account, it has been found to be counterproductive and resource inefficient.

What does all this evidence mean in terms of "feeding the world"? It becomes clear that industrial breeding has actually reduced food security by destroying small farms and the small farmers' capacity to produce these diverse outputs of nutritious crops. From the point of view of both food productivity and food entitlements, industrial agriculture is deficient compared to diversity-based internal input systems. Protecting small farms that conserve biodiversity is thus a food security imperative.

Data show that everywhere in the world, biodiverse small farms produce more agricultural output per unit area than large farms. Even in the United States, small farms of twenty-seven acres or fewer have ten times greater dollar output per acre than larger farms. It is therefore time to switch from measuring monoculture yields to assessing biodiversity outputs in farming systems.

Thus, both at the level of individual peasant farms and at the national level, the Green Revolution has led to a decline in food security. The same applies to the Gene Revolution. What the Green Revolution achieved was an increase in industrial inputs which, of course, created growth for the agrochemical and fossil fuel industries. But this increased consumption of toxins and energy by the agricultural sector did not translate into more food.

Today, most of the 1 billion people who lack adequate access to food are rural communities whose entitlements have collapsed, due either to environmental degradation or to livelihood destruction and negative terms of trade. Food security is therefore intimately connected to the livelihood security of small rural producers. There are proven alternatives to industrial agriculture and genetic engineering, and these are based on small farms and ecological methods. Sound resource use combined with social justice is the path of sustainability in agriculture that we should be taking.

The higher productivity of diversity-based systems indicates that there is an alternative to genetic engineering and industrial agriculture— an alternative that is more ecological and more equitable. This alternative is based on the intensification of biodiversity—intensifying through integrating diverse species—in place of chemical intensification, which promotes monocultures and, unlike its ecological alternative, fails to take all outputs of all species into account.

As Navdanya's work on biodiversity-based organic farming shows, India could feed twice its population through biodiversity intensification.[18] The U.N. report submitted to the General Assembly on December 20, 2010, also confirms that ecological agriculture produces more food: "Resource conserving, low-external-input techniques have a proven potential to significantly improve yields. Ecological interventions on 12.6 million farms increased crop yields of 79 percent."[19] A UNCTAD-UNEP study found that ecological methods increase crop yields by 116 percent for all of Africa and 128 percent in East Africa.[20]

Shifts for the Transition to the New Paradigm

The main shifts needed for a transition to a new agriculture paradigm are the following:

1. A shift from a reductionist, mechanistic paradigm of agricultural education, research, and extension to the holistic paradigm of agroecology.
2. A shift from an agriculture based on a war paradigm to one based on peace.
3. A shift from agricultural subsidies to chemical inputs to support to promote organic farming through training and facilitating access to markets from local to international levels.
4. A shift from chemical intensification to ecological intensification through intensifying biodiversity. Chemical-intensive agricultures uses more land and more resources, while ecologically intensive agriculture produces more nutrition while using less resources.
5. A shift from a focus on monocultures for commodity production based on subsidized external inputs to a multifunctional agriculture whose aim is maintaining and enriching nature's and people's economies, protection of biological and cultural diversity, maintaining the well-being of rural communities, creation of rural livelihoods, and production of high-quality nutritious food.
6. A shift from the reductionist measure of "yield" of commodities to the holistic measure of biodiverse outputs and multifunctional benefits through "health per acre" and "wealth per acre."
7. A shift from quantity to quality in measuring output of agriculture.
8. A shift from treating farmers and peasants as disposable and dispensable to recognizing their central role in maintaining ecosystems, cultures, and local economies. Farmers must be guaranteed respect, dignity, fair returns, and democratic participation.

A corporatized, industrialized, globalized farm system has given us hunger and malnutrition. We need to make a transition to people-centered, ecological, and decentralized food systems to address the deepening crisis of malnutrition and hunger. This transition involves radical changes in how food is produced and how it is distributed. A production system designed to end hunger and malnutrition has been put into practice on hundreds of thousands of farms.

As noted at the beginning of this essay, four hundred scientists who worked on the International Assessment of Agricultural Knowledge, Science and Technology report *Agriculture at the Crossroads,* sponsored by UNDP, FAO, UNEP, UNESCO, the World Bank, WHO, and Global Environmental Facility, clearly stated, "Business as usual is no longer an option." There needs to be a shift in the agricultural knowledge, science, and technology (AKST) paradigm from fragmented, component-based technologies to holistic, systems-based approaches. IAASTD moves away from single-commodity-based production systems to multifunctionality, which "recognizes the inescapable interconnectedness of agricultures' different roles and functions. The concept of multi-functionality recognizes agriculture as a multi output activity producing not only commodities (food, feed, fibres, agrofuels, medicinal products and ornamentals) but also non-commodity outputs such as environmental services, landscape amenities and cultural heritages." I would add to the definition of noncommodity output the biodiversity of nutritious foods produced for households and local economies.

The IAASTD recognizes that through an agroecological approach, "agro-ecosystems of even the poorest societies have the potential through ecological agriculture and IPM [integrated pest management] to meet or significantly exceed yields produced by conventional methods, reduce the demand for land conversion for agriculture, restore ecosystem services (particularly water) reduce the use of and need for synthetic fertilizers derived from fossil fuels, and the use of harsh insecticides and herbicides."

Navdanya's research and practice also shows that an ecological approach to agriculture delivers higher benefits in terms of food security than does industrial agriculture. Diversity goes hand in hand with decentralization. And the creation of decentralized biodiverse food systems is central to the design of a world without hunger. For this, a shift from globalization to localization is vital. Globalization has reduced food to a commodity while expanding the control of agribusiness over our food. Localization reclaims food as nourishment and expands community control over food systems.

Globalization	Localization
Agriculture and food systems shaped and controlled by corporations	Agriculture and food systems shaped and controlled by communities
Based on chemicals and GMOs, which bring profits for corporations	Based on biodiversity and agroecology, which bring benefits to ecosystems
Seed as intellectual property of corporations	Seed as common property of communities
Monocultures of a few commodities	Biodiversity of plants, animals, trees, soil organisms
Food as a commodity	Food as nourishment, food as a human right
Commodity speculation drives prices	Prices fixed by norms of justice and fairness
Hunger for 1 billion; food-related diseases for 2 billion	End of hunger and malnutrition—good food for all
Food dictatorship	Food democracy, food sovereignty

It is urgent that we design a transition from the globalization paradigm to a localization paradigm. This does not mean an end to international trade. But it does mean prioritizing the local. It means the decommodification of food, the reclamation of food as our being, our nourishment, our identity, our human right. It means freeing agriculture from WTO rules and governing it on the principles of food sovereignty. It means removing from our food system the gamblers who created "nuclear waste" and "toxic waste" on the balance sheets of investment firms before they bring down the food economy as they brought down the financial economy. It means stopping land grabs and the diversion of food for the poor to fuel for the cars of the rich. It means remembering that "everything is food" and "we are what we eat"—at the biological level, food justice is an ecological imperative. As biological beings, we all intrinsically have an equal share in the earth's resources and in their potential to provide food for us all. Seed grabs, land grabs, and food grabs violate the very ethical and ecological design of our being human. Hunger by design is immoral, unjust, and nonsustainable. We are capable of making a transition to a better design that is ethical, just, and sustainable.

Food wars are destroying the planet, our farmers, and our health while denying billions their right to food. Food peace is achievable. It is imperative that we make peace with Mother Earth by protecting our soil, seeds, and biodiversity, our water and climate. Food peace is necessary to protect our small farmers and our health. Food peace can ensure the food rights of all. Let us put our collective creative energies toward designing a future of food that protects the planet and brings abundant and good food to the last child, the last woman, the last person, the last being.

Tagore invites us to return to the soil to make peace with the earth:

> Let us all return to the soil
> That lays the corners of its garments
> And waits for us.
> Life rears itself from her breast,
> Flowers bloom from her smiles
> Her call is the sweetest music;
> Her lap stretches from one corner to the other,
> She controls the strings of life.
> Her warbling waters bring
> The murmur of life from all eternity.

Notes

1. Vandana Shiva, "Science and Politics in the Green Revolution," in *The Violence of the Green Revolution: Third World Agriculture, Ecology, and Politics* (London: Zed Books, 1991); Vandana Shiva, "Food and Water," in *Water Wars: Privatization, Pollution and Profit* (Cambridge, MA: South End, 2002).

2. Vandana Shiva, *Soil, Not Oil: Environmental Justice in a Time of Climate Crisis* (Cambridge, MA: South End, 2008).

3. *Why Is Every 4th Indian Hungry?* (India: Navdanya, 2001).

4. *Biodiversity Based Productivity: A New Paradigm for Food Security* (India: Navdanya, 2009).

5. Babasaheb R. Sonawane, *Chemical Contaminants in Human Milk: An Overview* (Washington, DC: U.S. Environmental Protection Agency, 1995).

6. K. Park, *Park's Textbook of Preventive and Social Medicine,* 21st ed. (India: Banarsidas Bhano, 2011).

7. D. Heber, "Vegetables, Fruits, and Phytoestrogens in the Prevention of Diseases," *Journal of Postgraduate Medicine* 50 (2004): 145–49.

8. Ibid.; "Health-Promoting Properties of Common Herbs," *American Journal of Clinical Nutrition* 70, no. 3 (1999): 491S–9S; "Flavonoid Rich Fraction from Sageretia Theezans Leaves Scavenges Reactive Oxygen Radical Species and Increases the Resistance of Low Density Lipoproteins to Oxidation," *Journal of Medicinal Food* 12, no. 6 (2009): 1310–15; Penny M. Kris-Etherton et al., "Bioac-

tive Compounds in Foods: Their Role in the Prevention of Cardiovascular Disease and Cancer," *American Journal of Medicine* 113 (2002): 71S–88S; Rui Hai Liu, "Potential Synergy of Phytochemicals in Cancer Prevention: Mechanism of Action," *Journal of Nutrition* 134, no. 12 (2004).

9. Liu, "Potential Synergy of Phytochemicals in Cancer Prevention"; *Diet and Lifestyle Recommendations, Revision 2006—A Scientific Statement from the American Heart Association Nutrition Committee.*

10. Kris-Etherton et al., "Bioactive Compounds in Foods"; Liu, "Potential Synergy of Phytochemicals in Cancer Prevention"; John W. Finley, "Proposed Criteria for Assessing the Efficacy of Cancer Reduction by Plant Foods Enriched in Carotenoids, Glucosinolates, Polyphenols and Selenocompounds," *Oxford Journals, Life Sciences, Annals of Botany* 95, no. 7 (2005); *Diet and Lifestyle Recommendations Revision 2006—A Scientific Statement from the American Heart Association Nutrition Committee.*

11. Park, *Park's Textbook.*

12. Ibid.

13. Ibid.

14. Ibid.

15. Virginia Worthington, "Nutritional Quality of Organic versus Conventional Fruits, Vegetables, and Grains," *Journal of Alternative and Complementary Medicine* 7, no. 2 (2001).

16. *Biodiversity Based Productivity: A New Paradigm for Food Security* (New Delhi: Navdanya, 2009).

17. Ibid.; *Health per Acre* (India: Navdanya, 2011).

18. *Health per Acre.*

19. Olivier de Schutter, report submitted to the Special Rapporteur on the Right to Food, December 2010, 8.

20. UNCTAD-UNEP, *Organic Agriculture and Food Security in Africa* (New York and Geneva: United Nations, 2008), 16.

7

Can Life Be Made?
Can Life Be Owned?

Redefining Biodiversity

In 1971, General Electric and one of its employees, Anand Mohan Chakravarty, applied for a U.S. patent on a genetically engineered pseudomonas bacteria. Taking plasmids from three kinds of bacteria, Chakravarty transplanted them into a fourth. As he explained, "I simply shuffled genes, changing bacteria that already existed." Chakravarty was granted his patent on the grounds that the microorganism was not a product of nature but his invention and, therefore, patentable. As Andrew Kimbrell, a leading U.S. lawyer, recounts, "In coming to its precedent-shattering decision, the court seemed unaware that the inventor himself had characterized his 'creation' of the microbe as simply 'shifting' genes, not creating life."[1]

On such slippery grounds, the first patent on life was granted, and in spite of the exclusion of plants and animals from patenting under U.S. law, the United States has since rushed to grant patents on all kinds of life-forms.

Currently, well over 190 genetically engineered animals, including fish, cows, mice, and pigs, are figuratively standing in line to be patented by a variety of researchers and corporations. According to Kimbrell: "The Supreme Court's Chakravarty decision has been extended to be continued, up the chain of life. The patenting of microbes has led inexorably to the patenting of plants, and then animals."[2]

Biodiversity has been redefined as "biotechnological inventions" to make the patenting of life-forms appear less controversial. These patents are valid for twenty years and hence cover future generations of plants and animals. Yet even when scientists in universities or corporations shuffle genes, they do not "create" the organism that they then patent. Referring to the landmark Chakravarty case, in which the court found that he had "produced a new bacterium with markedly different characteristics than any found in nature," Key Dismukes, study director for the Committee on Vision of the National Academy of Sciences in the United States, said:

> Let us at least get one thing straight: Anand Chakravarty did not create a new form of life; he merely intervened in the normal processes by which strains of bacteria exchange genetic information, to produce a new strain with an altered metabolic pattern. "His" bacterium lives and reproduces itself under the forces that guide all cellular life. Recent advances in recombinant DNA techniques allow more direct biochemical manipulation of bacterial genes than Chakravarty employed, but these too are only modulations of biological processes. We are incalculably far away from being able to create life de novo, and for that I am profoundly grateful. The argument that the bacterium is Chakravarty's handiwork and not nature's wildly exaggerates human power and displays the same hubris and ignorance of biology that have had such devastating impact on the ecology of our planet.[3]

This display of hubris and ignorance becomes even more conspicuous when the reductionist biologists who claim patents on life declare that 95 percent of DNA is "junk DNA," meaning that its function is not known. When genetic engineers claim to "engineer" life, they often have to use this "junk DNA" to get their results.

Take the case of a sheep named Tracy, a "biotechnological invention" of the scientists of Pharmaceutical Proteins Ltd. (PPL). Tracy is called a "mammalian cell bioreactor" because, through the introduction of human genes, her mammary glands are engineered to produce a protein, alpha-1-antitrypsin, for the pharmaceutical industry. As Ron James, director of PPL, states, "The mammary gland is a very good factory. Our sheep are furry little factories walking around in fields and they do a superb job." While they claim that genetic engineers created the "biotechnological invention," the scientists at PPL had to use "junk DNA" to get high yields of alpha-1-antitrypsin. According to James, "We left some of these random bits of DNA in the gene, essentially as God provided it and that produced high yield." In claiming the patent, however, it is the scientist who becomes God, the creator of the patented organism.

Further, future generations of the animal are clearly not "inventions" of the patent holder; they are the product of the regenerative capacity of the organism. Thus, though the metaphor for patenting is "engineers" who "make machines," of the 550 sheep eggs injected with hybrid DNA, 499 survived. When these were transplanted into surrogate mothers, only 112 lambs were born, just 5 of which had incorporated the human gene into their DNA. Of these, only 3 produced alpha-1-antitrypsin in their milk, 2 of which delivered three grams of proteins per liter of milk. But Tracy is the only lamb among the 112 engineered ones to become PPL's "sheep that lays golden eggs" and produce thirty grams per liter.

One of the characteristics of reductionist biology is to declare organisms and their functions useless on the basis of ignorance of their structure and function. Thus, crops and trees are declared "weeds."[4] Forests and cattle breeds are declared "scrub." And DNA whose role is not understood is called "junk DNA." To write off the major part of the molecule as junk because of our ignorance is to fail to understand biological processes. "Junk DNA" plays an essential role. The fact that Tracy's protein production increased with the introduction of "junk DNA" is an illustration of the PPL scientists' ignorance, not their knowledge and creativity.

While genetic engineering is modeled on determinism and predictability, indeterminism and unpredictability are characteristic of the human manipulation of living organisms. In addition to the gap between the projection and practice of the engineering paradigm, there is the gap between owning benefits and rewards and owning hazards and risks.

When property rights to life-forms are claimed, it is on the basis of their being new, novel, not occurring in nature. But when it comes time for the "owners" to take responsibility for the consequences of releasing genetically modified organisms, suddenly the life-forms are not new. They are natural, and hence safe. The issue of biosafety is treated as unnecessary.[5] Thus, when biological organisms have to be owned, they are treated as not natural; when the ecological impact of releasing GMOs is called to account by environmentalists, these same organisms are now natural. These shifting constructions of "natural" show that science, which claims the highest levels of objectivity, is actually very subjective and opportunistic in its approach to nature.

The inconsistency in the construction of "natural" is well illustrated in the case of the manufacture of genetically engineered human proteins for infant formula. Gen Pharm, a biotechnology company, is the owner of the world's first transgenic dairy bull, called Herman. Herman was bioengineered by company scientists when still an embryo to carry a human gene for producing milk with a human protein. The milk was to be used for making infant formula.

The engineered gene and the organism of which it is a part are treated as nonnatural when it comes to ownership of Herman and his offspring. Yet when it comes to the safety of the infant formula containing this bioengineered ingredient extracted from the udders of Herman's offspring, the same company says, "We're making these proteins exactly the way they're made in nature." Gen Pharm's chief executive officer, Jonathan MacQuitty, would have us believe that infant formula made from human protein bioengineered in the milk of transgenic dairy cattle is human milk. "Human milk is the gold standard, and formula companies have added more and more [human elements] over the past 20 years." From this perspective, cows, women, and children are merely instruments for commodity production and profit maximization.[6]

As though the inconsistency between the construction of the natural and novel in the spheres of patent protection and health and environmental protection was not enough, Gen Pharm, the "owner" of Herman, has totally changed the objective for making a transgenic bull. They now have ethical clearance on the grounds that, by using him for breeding, the modified version of the human gene for lactoferin might be of benefit to patients with cancer or AIDS.

Patenting living organisms encourages two forms of violence. First, life-forms are treated as if they are mere machines, thus denying their self-organizing capacity. Second, by allowing the patenting of future generations of plants and animals, the self-reproducing capacity of living organisms is denied. Living organisms, unlike machines, organize themselves. Because of this capacity, they cannot be treated as simply "biotechnological inventions," "gene constructs," or "products of the mind" that need to be protected as "intellectual property." The engineering paradigm of biotechnology is based on the assumption that life can be made. Patents on life are based on the assumption that life can be owned because it has been constructed.

Genetic engineering and patents on life are the ultimate expression of the commercialization of science and the commodification of nature that began the scientific and industrial revolutions. As Carolyn Merchant has analyzed in *The Death of Nature,* the rise of reductionist science allowed nature to be declared dead, inert, and valueless. Hence, it allowed for the exploitation and domination of nature, in total disregard of the social and ecological consequences.[7] The rise of reductionist science was linked with the commercialization of science, and resulted in the domination of women and non-Western peoples. Their diverse knowledge systems were not treated as legitimate ways of knowing. With commercialization as the objective, reductionism became the criterion of scientific validity. Nonreductionist and ecological ways of knowing, and nonreductionist and ecological systems of knowledge, were pushed out and marginalized.

The genetic engineering paradigm is now pushing out the last remains of ecological paradigms by redefining living organisms and biodiversity as "man-made" phenomena. The rise of the reductionist paradigm of biology to serve the commercial interests of the genetic engineering, biotechnology industry was itself engineered. This was done through funding as well as rewards and recognition.

Genetic Engineering and the Rise of the Reductionist Paradigm of Biology

Reductionism in biology is multifaceted. At the species level, this reductionism puts value on only one species—humans—and generates an instrumental value for all others. It therefore displaces and pushes to extinction all species that have no or low instrumental value to humans. Monocultures of species and biodiversity erosion are the inevitable consequences of reductionist thought in biology, especially when applied to forestry, agriculture, and fisheries. We call this first-order reductionism.

Reductionist biology is increasingly characterized by a second-order reductionism—genetic reductionism—the reduction of all behavior of biological organisms, including humans, to genes. Second-order reductionism amplifies the ecological risks of first-order reductionism, while introducing new issues, like the patenting of life-forms.

Reductionist biology is also an expression of cultural reductionism, since it devalues many forms of knowledge and ethical systems. This includes all non-Western systems of agriculture and medicine as well as all disciplines in Western biology that do not lend themselves to genetic and molecular reductionism, but are necessary for dealing sustainably with the living world.

Reductionism was promoted strongly by August Weismann, who nearly a century ago postulated the complete separation of the reproductive cells—the germ line—from the functional body, or soma. According to Weismann, reproductive cells are already set apart in the early embryo and continue their segregated existence into maturity, when they contribute to the formation of the next generation. This supported the idea that acquired traits with no direct feedback from the environment were noninheritable. The mostly nonexistent "Weismann barrier" is still the paradigm used to discuss biodiversity conservation as "germ plasm" conservation. The germ plasm, Weismann had earlier contended, was divorced from the outside world. Evolutionary changes toward greater fitness—meaning greater capacity to reproduce—were the result of fortuitous mistakes that happened to prosper in the competition of life.[8]

Weismann's classic experiment a century ago was taken as proof of the noninheritability of acquired characteristics. He cut the tails off twenty-two generations of mice and found that the next generation was still born with normal tails. The sacrifice of hundreds of mouse tails only proved that this type of mutilation was not inherited.[9]

The proposition that information only goes from genes to the body was reinforced by molecular biology and the discovery in the 1950s of the role of nucleic acid, placing Mendelian genetics on a solid material basis. Molecular biology showed a means of transferal of information from genes to proteins, but gave no indication—until recently—of any transfer in the opposite direction. The inference that there could be none became what Francis Crick called the central dogma of molecular biology: "Once 'information' has passed into proteins, it cannot get out again."[10]

Isolating the gene as a "master molecule" is part of biological determinism. The "central dogma" that genes as DNA make proteins is another aspect of this determinism. This dogma is preserved even though it is known that genes "make" nothing. As Richard Lewontin states in *The Doctrine of DNA*:

> DNA is a dead molecule, among the most non-reactive, chemically inert molecules in the world. It has no power to reproduce itself. Rather, it is produced out of elementary materials by a complex cellular machinery of proteins. While it is often said that DNA produces proteins, in fact proteins (enzymes) produce DNA.
>
> When we refer to genes as self-replicating, we endow them with a mysterious autonomous power that seems to place them above the more ordinary materials of the body. Yet if anything in the world can be said to be self-replicating, it is not the gene, but the entire organism as a complex system.[11]

Genetic engineering is taking us into a second-order reductionism not only because organisms are perceived in isolation of their environment, but because genes are perceived in isolation of the organism as a whole. The doctrine of molecular biology is modeled on classical mechanics. The central dogma is the ultimate in reductionist thought.

At the very same time that Max Planck, Niels Bohr, Albert Einstein, Erwin Schrodinger, and their brilliant colleagues were revising the Newtonian view of the physical universe, biology was becoming more reductionist.[12] Reductionism in biology was not an accident but a carefully planned paradigm. As Lily E. Kay records in *The Molecular Vision of Life*, the Rockefeller Foundation served as a principal patron of molecular biology from the 1930s to the 1950s. The term *molecular biology* was coined in 1938 by Warren Weaver, the director of the Rockefeller Foundation's Natural Science Division. The term was intended to capture the essence of the foundation's program—its emphasis on the ultimate minuteness of biological entities. The cognitive and structural reconfigurations of biology into a reductionist paradigm were greatly facilitated through the economically powerful Rockefeller Foundation. During the years 1932–1959, the foundation poured about $25 million into molecular biology programs in the United States, more than one-fourth of the foundation's total spending for the biological sciences outside of medicine (including, from the early 1940s on, enormous sums for agriculture).[13] The force of the foundation's funding set the trends in molecular biology. During the dozen years following 1953 (the elucidation of the structure of DNA), Nobel Prizes were awarded to scholars for research into the molecular biology of the gene, and all but one had been either fully or partially sponsored by the Rockefeller Foundation under Weaver's guidance.[14]

The motivation behind the enormous investment in the new agenda was to develop the human sciences as a comprehensive explanatory and applied framework of social control grounded in the natural, medical, and social sciences. Conceived during the late 1920s, the new agenda was articulated in terms of the contemporary technocratic discourse of human engineering, aiming toward restructuring human relations in congruence with the social framework of industrial capitalism. Within that agenda, the new biology (originally named "psychobiology") was erected on the bedrock of the physical sciences in order to rigorously explain and eventually control the fundamental mechanisms governing human behavior, placing a particularly strong emphasis on heredity. Hierarchy and inequality were thus "naturalized." As Lewontin states in *The Doctrine of DNA:* "The naturalistic explanation is to say that not only do we differ in our innate capacities but that these innate capacities are themselves transmitted from generation to generation biologically. That is to say, they are in our genes. The original social and economic notion of inheritance has been turned into biological inheritance."[15]

The conjunction of cognitive and social goals in reductionist biology had a strong historical connection to eugenics. As of 1930, the Rockefeller Foundation had supported a number of eugenically directed projects. By the time the "new science of man" was inaugurated, however, the goal of social control through selective breeding was no longer socially legitimate.

Precisely because the old eugenics had lost its scientific validity, a space was created for a new program that promised to place the study of human heredity and behavior on vigorous grounds. A concerted physicochemical attack on the gene was initiated at the moment in history when it became unacceptable to advocate social control based on crude eugenic principles and outmoded racial theories. The molecular biology program, through the study of simple biological systems and the analyses of protein structure, promised a surer, albeit much slower, way toward social planning based on sounder principles of eugenic selection.[16]

Reductionism was chosen as the preferred paradigm for economic and political control of diversity in nature and society. Genetic determinism and genetic reductionism go hand in hand. But to say that genes are primary is more ideology than science. Genes are not independent entities but dependent parts of an entirety that gives them effect. All parts of the cell interact, and the combinations of genes are at least as important as their individual effects in the making of an organism. More broadly, an organism cannot be treated simply as the product of a number of proteins, each produced by the corresponding gene. Genes have multiple effects, and most traits depend on multiple genes.

Yet the linear and reductionist causality of genetic determinism is held on to, even though the very processes that make genetic engineering possible run counter to the concepts of "master molecules" and the "central dogma." As Roger Lewin has stressed: "Restriction sites, promoters, operators, operons, and enhancers play their part. Not only does DNA make RNA, but RNA, aided by an enzyme suitably called reverse transcriptase, makes DNA."[17]

The weakness of the explanatory and theoretical power of reductionism is made up for by its ideological power as well as its economic and political backing. Some biologists have gone far in exalting the gene over the organism and demoting the organism itself to a mere machine. The sole purpose of this machine is its own survival and reproduction or, perhaps more accurately put, the survival and reproduction of the DNA that is said both to program and to "dictate" its operation. In Richard Dawkins's terms, an organism is a "survival machine"—a "lumbering robot" constructed to house its genes, those "engines of self-preservation" that have as their primary property inherent "selfishness." They are sealed off from the outside world, communicating with it by tortuously indirect routes, manipulating it by remote control. They are in you and in me; they created us, body and mind. And their preservation is the ultimate rationale for our existence.[18]

This reductionism has epistemological, ethical, ecological, and socioeconomic implications. Epistemologically, it leads to a machine view of the world and its rich diversity of life-forms. It makes us forget that living organisms organize themselves. It robs us of our capacity for the reverence for life—and without that capacity, protection of the diverse species on this planet is impossible.

Engineering versus Growing

The capacity to self-organize is the distinctive feature of living systems. Self-organizing systems are autonomous and self-referential. This does not mean that they are isolated and noninteractive. Self-organized systems interact with their environment but maintain their autonomy. The environment merely triggers the structural changes; it does not specify or direct them. The living system specifies its own structural changes and which patterns in the environment will trigger them. A self-organizing system knows what it has to import and export in order to maintain and renew itself.

Living systems are also complex. The complexity of their structure allows for self-ordering and self-organization. It also allows for the emergence of new properties. One of the distinguishing properties of living systems is their ability to undergo continual structural changes while preserving their form and pattern of organization.

Living systems are also diverse. Their diversity and uniqueness are maintained through spontaneous self-organization. The components of a living system are continually renewed and recycled with structural interaction with the environment, yet the system maintains its pattern, its organization, and its distinctive form.

Self-healing and repair is another characteristic of living systems that derives from complexity and self-organization.

The freedom for diverse species and ecosystems to self-organize is the basis of ecology. Ecological stability derives from the ability of species and ecosystems to adapt, evolve, and respond. In fact, the more degrees of freedom available to a system, the more a system can express its self-organization.

External control reduces the degrees of freedom a system has, thereby reducing its capacity to organize and renew itself. Ecological vulnerability comes from the fact that species and ecosystems have been engineered and controlled to such an extent that they lose the capacity to adapt and evolve.

Chilean scientists Humberto R. Maturana and Francisco J. Varela have distinguished two kinds of systems—autopoietic and allopoietic. A system is autopoietic when its function is primarily geared toward self-renewal. An autopoietic system refers to itself. In contrast, an allopoietic system, such as a machine, refers to a function given from outside, such as the production of a specific output.[19]

Self-organizing systems grow from within, shaping themselves outward. Externally organized mechanical systems do not grow; they are made, put together from the outside. Self-organizing systems are distinct and multidimensional. They therefore display structural and functional diversity. Mechanical systems are uniform and unidimensional. They display structural uniformity and functional one-dimensionality. Self-organizing systems can heal themselves and adapt to changing environmental conditions. Mechanically organized systems do not heal or adapt; they break down.

The more complex a dynamic structure is, the more endogenously it is driven. Change depends not only on its external compulsions but on its internal conditions. Self-organization is the essence of health and ecological stability for living systems.

When an organism or a system is mechanically manipulated to improve a one-dimensional function, including the increase in one-dimensional productivity, either the organism's immunity decreases and it becomes vulnerable to disease and attack by other organisms, or the organism becomes dominant in an ecosystem and displaces other species, pushing them into extinction. Ecological problems arise from applying the engineering paradigm to life. This paradigm is being deepened through genetic engineering, which will have major ecological and ethical implications.

Ethical Implications of Genetic Engineering

When organisms are treated as if they are machines, an ethical shift takes place—life is seen as having instrumental rather than intrinsic value. The manipulation of animals for industrial ends has already had major ethical, ecological, and health implications. The reductionist, machine view of animals removes all barriers of ethical concern for how animals are treated to maximize production. Within the industrial livestock production sector, the mechanistic view predominates. For example, a manager of the meat industry states that: "The breeding sow should be thought of as, and treated as, a valuable piece of machinery, whose function is to pump out baby pigs like a sausage machine."[20]

Treating pigs as machines, however, has a major impact on their behavior and health. In animal factories, pigs have to have their tails, teeth, and testicles cut off because they fight with each other and resort to what the industry calls "cannibalism." Eighteen percent of the piglets in factory farms are choked to death by their mother. Two to five percent are born with congenital defects, such as splayed legs, no anus, or inverted mammary glands. They are prone to diseases such as "banana disease" (so named because stricken pigs arch their backs into a banana shape) or porcine stress syndrome. These stresses and diseases are bound to increase with genetic engineering. Already, the pig with the human growth hormone has a body weight that is more than its legs can carry.

The issues of health and animal welfare are intrinsically related to the ecological impact of the new technologies on the capacity of self-regulation and healing. The issue of intrinsic worth is intimately related to the issue of self-organization, which is also, in turn, related to healing.

In the making of the organism, the multiplying cells seem to be instructed as to their respective destinies, and they become permanently differentiated to compose organs. But the instructions or pattern for making the whole structure remain somehow latent. When a part is injured, some cells become undifferentiated in order to make new, specialized tissues.[21]

Thus, there is a self-directed capacity for restoration. The faculty of repair is, in turn, related to resilience. When organisms are treated as machines, and manipulated without recognition of their ability to self-organize, their capacity to heal and repair breaks down, and they need increasing inputs and controls to be maintained.

Ecological and Socioeconomic Implications of Genetic Engineering

Genetic engineering has epistemological and ethical implications not merely for the material conditions of our life, our health, and our environment. Health implications are built into the very techniques of genetic engineering.

Genetic engineering moves genes across species by using "vectors"— usually a mosaic recombination of natural genetic parasites from different sources, including viruses causing cancers and other diseases in animals and plants that are tagged with one or more antibiotic-resistant "marker" genes. Evidence accumulating over the past few years confirms the fear that these vectors constitute major sources of genetic pollution with drastic ecological and public health consequences. Vector-mediated horizontal gene transfer and recombination are found to be involved in generating new pandemic strains of bacterial pathogens.[22]

Genetic engineering also has major ecological impacts, even though the biotechnology industry and regulatory agencies keep claiming that there have been no adverse consequences from the over five hundred field releases in the United States.[23] Existing field tests are not designed to collect environmental data, and test conditions do not approximate production conditions that include commercial scale, varying environments, and time periods. Yet, as Phil J. Regal has stated, "This sort of nondata on nonreleases has been cited in policy circles as though 500 true releases have now informed scientists that there are no legitimate scientific concerns."[24]

Two studies of detailed environmental impact assessment have verified the hazards posed by the large-scale introduction of genetically engineered organisms in the field of agriculture. At the 1994 annual meeting of the Ecological Society of America, researchers from Oregon State University reported on tests to evaluate a genetically engineered bacterium designed to convert crop waste into ethanol. A typical root zone–inhabiting bacterium, *Klebsiella planti-cola,* was engineered with the novel ability to produce ethanol, and the engineered bacterium was added to enclosed soil chambers in which a wheat plant was growing. In one soil type, all the plants in soil with the engineered bacterium died, while plants in untreated soil remained healthy.

In all cases, mycorrhizal fungi in the root system were reduced by more than half, which ruined nutrient uptake and plant growth. This result was unpredicted. Reduction in this vital fungus is known to result in plants that are less competitive with weeds or more susceptible to disease. In low-organic-matter sandy soil, the plants died from ethanol produced by the engineered bacterium in the root system, while in high-organic-matter sandy or clay soil, changes in nematode density and species composition resulted in significantly decreased plant growth. The lead researcher, Dr. Elaine Ingham, concluded that these results imply that there can be significant and serious effects resulting from the addition of a genetically engineered microorganism (GEM) to soil. The tests, using a new and comprehensive system, disproved earlier suggestions that there were no significant ecological effects.[25]

In 1994, research scientists in Denmark reported strong evidence that an oilseed rape plant genetically engineered to be herbicide tolerant transmitted its transgene to a weedy natural relative, *Brassica campestris* ssp. *campestris*. This transfer can take place in just two generations of the plant. In Denmark, *B. campestris* is a common weed in cultivated oilseed rape fields, where selective elimination by herbicides is now impossible. The wild relative of this weed is spread over large parts of the world. One way to assess the risk of releasing transgenic oilseed rape is to measure the rate of natural hybridization with *B. campestris,* because certain transgenes could make its wild relative a more aggressive weed, even harder to control.

Although crosses with *B. campestris* have been used in the breeding of oilseed rape, natural interspecific crosses with oilseed rape were generally thought to be rare. Artificial crosses by hand pollination carried out in a risk-assessment project in the United Kingdom were reported to be unsuccessful. A few studies, however, have reported spontaneous hybridization between oilseed rape and the parented species *B. campestris* in field experiments. As early as 1962, hybridization rates of 0.3 to 88 percent were measured for oilseed rape and wild *B. campestris.* The results of the Danish team showed that high levels of hybridization can occur in the field. Field tests revealed that between 9 and 93 percent of hybrid seeds were produced under different conditions.[26]

The transfer of herbicide resistance to wild, weedy relatives of crops threatens to create "superweeds" that are resistant to herbicides and hence uncontrollable. As a strategy for Monsanto to sell more Roundup and Ciba Geigy to sell more Basta, genetically engineered herbicide-resistant crops make sense. Yet this strategy runs counter to a policy of sustainable agriculture, since it undermines the very possibility of weed control.

Just as the strategy of using genetic engineering for herbicide resistance fails to control weeds and instead carries the risk of creating "superweeds," the strategy of genetically engineered crops for pest resistance fails to control pests and instead carries the risk of creating "superpests." In 1996, nearly 2 million acres in the United States were planted with a genetically engineered cotton variety from Monsanto called Bollgard. Monsanto's Bollgard cotton is a transgenic variety that has been engineered with DNA from the soil microbe *Bacillus thurengesis* (Bt) to produce proteins poisonous to the bollworm, a cotton pest. Monsanto charged the farmers a "technology fee" of $79 per hectare in addition to the price of seed for "peace of mind" through "seasonlong plant control . . . that stops worm problems before they start." The company collected $51 million in one year alone from this "technology fee."[27]

The technology, however, has already failed the farmers. The bollworm infestation on the genetically engineered crop was more than twenty to fifty times the level that typically triggers spraying. Further, since Bt has been an important natural biological control agent used by organic farmers, the genetic engineering strategy undermines the organic strategy.[28]

Besides the "technology fee," Monsanto has also placed highly restrictive rules on farmers. As the company states: "Monsanto is only licensing growers to use seed containing the patented Bollgard gene for one crop. Saving or selling the seed for replanting will violate the limited license and infringe upon the patent rights of Monsanto. This may subject you to prosecution under federal law."[29] Monsanto "owns" the crop when it comes to reaping millions of dollars in rent from farmers, but it does not own the costs or take responsibility for the hazards that its transgenic crop generates.

Intellectual property rights (IPR) monopolies are justified on grounds that corporations are given IPRs by society so that society can benefit from their contributions. The failure of the transgenic cotton shows that the assumption that IPRs will "improve" agriculture does not always hold.

Instead, what we have is an example of social and ecological costs generated for society in general and farmers in particular. IPRs on crop varieties that are creating ecological havoc is an unjust system of total privatization of benefits and total socialization of costs.

Monopolies linked to this unaccountable and unjust system prevent the development of ecologically sound and socially just practices. Further, they force an agricultural system on people that threatens the environment and human health.

The imposition of monopolies, and of genetically engineered products, is, ironically, at the core of the "free trade" system. Legally, it is a free trade treaty, the Uruguay Round of GATT, that is forcing all countries to have IPRs in agriculture. Economically, the introduction of genetically engineered products is being forced on unwilling citizens and countries on the grounds of "free trade" which, as the case of Monsanto's soybeans illustrates, translates into the absolute freedom of transnational corporations to force hazardous products on people.

World Food Day, October 16, 1996, was celebrated by five hundred organizations from seventy-five countries calling for an international boycott of genetically engineered soybeans resistant to the chemical herbicide glyphaosate, which Monsanto sells as Roundup. Monsanto had genetically engineered the soybean to increase its herbicide sale.[30]

This was also a major controversy at the World Food Summit held in Rome in November 1996. Monsanto, which claimed its soybean was distinctive and novel to get a patent, now says that the new soybean is indistinguishable from the conventional bean in order to mix the two types of soybeans together offshore and import them to European markets. Citizens are demanding that the genetically engineered soy be labeled under their "right to know" and "right to choose."

Both the soybean and cotton are now Monsanto monopolies since it acquired Agracetus, which has broad species patents for all transgenic cotton and soy, in May 1996 for $150 million. These patents are given on the basis of novelty, but that novelty is denied in the face of consumer resistance and concern over the safety of genetically engineered products.

As a technique, genetic engineering is very sophisticated. But as a technology for using biodiversity sustainably to meet human needs, it is clumsy. Transgenic crops reduce biodiversity by displacing diverse crops, which provide diverse sources of nutrition.

In addition, new health risks are being introduced through transgenic crops. Genetically engineered foods have the potential of introducing new allergies. They also carry the risk of "biological pollution," of new vulnerability to disease, of one species becoming dominant in an ecosystem, and of gene transfer from one species to another.

In an experiment carried out in the United Kingdom by Dr. James Bishop, scorpion genes were introduced into a virus to make an insecticidal spray to kill caterpillars. The transgenic virus is assumed to be safe on grounds that it will not cross species boundaries for its target, even though there are plenty of examples of viruses and disease organisms finding new target species. Scientific evidence also shows that genetic engineering can create "superviruses," viruses resistant to pesticides. Complacency on biosafety issues is therefore not justified on the basis of available scientific evidence.

A clearance has recently been given for the first trial of genetically engineered crops in India. They include a tomato engineered with Bt and hybrid brassica. There is already enough scientific evidence that genetic engineering with Bt is contributing to resistance and therefore is not a sustainable route for controlling plant pests and disease.

The promised benefits of genetically engineered crops and foods are illusionary, although their potential risks are real. The illusion of genetic engineering is, however, not merely at the systems level in food production and consumption. It is also at the scientific level. Genetic engineering offers its promises on the basis of genetic reductionism and determinism. Yet both of these assumptions are being proved false through molecular biology research itself.

Celebrating and Conserving Life

In the era of genetic engineering and patents, life itself is being colonized. Ecological action in the biotechnology era involves keeping the self-organization of living systems free—free of technological manipulations that destroy the self-healing and self-organizational capacity of organisms, and free of legal manipulations that destroy the capacities of communities to search for their own solutions to human problems from the richness of the biodiversity that we have been endowed with.

There are two strands in my current work that respond to the manipulation and monopolization of life. Through Navdanya, a national network for setting up community seed banks to protect indigenous seed diversity, we have tried to build an alternative to the engineering view of life. Through work to protect the intellectual commons—either in the form of seed satyagraha launched by the farmers' movement or in the form of the movement for common intellectual rights that we have launched with the Third World Network—we have tried to build an alternative to the paradigm of knowledge and life itself as private property.

It is this freedom of life and freedom to live that I increasingly see as the core element of the ecology movement as we reach the end of the millennium. And in this struggle, I frequently draw inspiration from the Palestinian poem "The Seed Keepers":

Burn our land
burn our dreams
pour acid onto our songs
cover with sawdust
the blood of our massacred people
muffle with your technology
the screams of all that is free,
wild and indigenous.
Destroy
Destroy
our grass and soil
raze to the ground
every farm and every village
our ancestors had built
every tree, every home
every book, every law
and all the equity and harmony.
Flatten with your bombs
every valley; erase with your edits
our past,
our literature, our metaphor
Denude the forests
and the earth
till no insect,

no bird
no word
can find a place to hide.
Do that and more.
I do not fear your tyranny
I do not despair ever
for I guard one seed
a little live seed
that I shall safeguard
and plant again.

Notes

1. Andrew Kimbrell, *The Human Body Shop* (New York: Harper-Collins, 1993).

2. Ibid.

3. Key Dismukes, quoted in Brian Belcher and Geoffrey Hawtin, *A Patent on Life: Ownership of Plant and Animal Research* (Canada: IDRC, 1991).

4. Vandana Shiva, *Monocultures of the Mind* (London: Zed Books, 1993).

5. Rural Development Foundation International Communique, Ontario, Canada, June 1993.

6. *New Scientist,* January 9, 1993.

7. Carolyn Merchant, *The Death of Nature: Women, Ecology and the Scientific Revolution* (New York: Harper and Row, 1980), 182.

8. Robert Wesson, *Beyond Natural Selection* (Cambridge, MA: MIT Press, 1993), 19.

9. J. W. Pollard, "Is Weismann's Barrier Absolute?" in *Beyond Neo-Darwinism: Introduction to the New Evolutionary Paradigm,* ed. M. W. Ho and P. T. Saunders (London: Academic, 1984), 291–315.

10. Francis Crick, "Lessons from Biology," *Natural History* 97 (November 1988): 109.

11. Richard Lewontin, *The Doctrine of DNA* (Penguin Books, 1993).

12. Wesson, *Beyond Natural Selection,* 29.

13. Lily E. Kay, *The Molecular Vision of Life: Caltech, the Rockefeller Foundation and the Rise of the New Biology* (Oxford: Oxford University Press, 1993), 6.

14. Ibid., 8.

15. Lewontin, *The Doctrine of DNA,* 22.

16. Kay, *The Molecular Vision of Life,* 8–9.

17. Roger Lewin, "How Mammalian RNA Returns to Its Genome," *Science* 219 (1983): 1052–54.

18. Richard Dawkins, *The Selfish Gene* (Oxford: Oxford University Press, 1976).

19. Humberto R. Maturana and Francisco J. Varela, *The Tree of Knowledge: The Biological Roots of Human Understanding* (Boston: Shambala, 1992).

20. L. J. Taylor, quoted in David Coats, *Old McDonald's Factory Farm* (New York: Continuum, 1989), 32.

21. Wesson, *Beyond Natural Selection.*

22. Mae Wan Ho, "Food, Facts, Fallacies and Fears" (paper presented at National Council of Women Symposium, United Kingdom, March 22, 1996).

23. Vandana Shiva et al., *Biosafety* (Penang: Third World Network, 1996).

24. Phil J. Regal, "Scientific Principles for Ecologically Based Risk Assessment of Transgene Organisms," *Molecular Biology* 3 (1994): 5–13.

25. Elaine Ingham and Michael Holmes, "A Note on Recent Findings on Genetic Engineering and Soil Organisms," 1995.

26. R. Jorgensen and B. Anderson, "Spontaneous Hybridization between Oil-seed Rape (Brassica Napas) and Weedy B. campestris (Brassicaceae): A Risk of Growing Genetically Modified Oilseed Rape," *American Journal of Botany* (1994).

27. Rural Development Foundation International Communique, United States, July/August 1996, 7–8.

28. "Pests Overwhelm Bt. Corron Crop," *Science* 273:423.

29. Rural Development Foundation International Communique, United States, July/August 1996, 7–8.

30. The Battle of the Bean, "Splice of Life" (October 1966).

8

The Seed and the Earth

Regeneration lies at the heart of life and has been the central principle guiding sustainable societies; without regeneration, there can be no sustainability. Modern industrial society, however, has no time for thinking about regeneration and therefore no space for living regeneratively. Its devaluation of the processes of regeneration is the cause of both the ecological crisis and the crisis of sustainability.

In the *Rig Veda,* the hymn to the healing plants, medicinal plants are referred to as mothers because they sustain us.

> Mothers, you have a hundred forms
> and a thousand growths.
> You who have a hundred ways of
> working, make this person
> whole for me.
> Be joyful, you plants that bear
> flowers and those that bear fruit.

The continuity between regeneration in human and nonhuman nature that was the basis of all ancient worldviews was broken by patriarchy. People were separated from nature, and the creativity involved in processes of regeneration was denied. Creativity became the monopoly of men, who were considered to be engaged in production; women were engaged in mere reproduction or recreation, which, rather than being treated as renewable production, was looked upon as nonproductive.

Activity, as purely male, was constructed on the separation of the earth from the seed, and on the association of an inert and empty earth with the passivity of the female. The symbols of the seed and the earth, therefore, undergo a metamorphosis when cast in a patriarchal mold;

gender relations as well as our perception of nature and its regeneration are also restructured. This nonecological view of nature and culture has formed the basis of patriarchal perceptions of gender roles in reproduction across religions and through the ages.

This gendered seed/earth metaphor has been applied to human production and reproduction to make the relationship of dominance of men over women appear natural. But the naturalness of this hierarchy is built on a material/spiritual dualism, with male characteristics artificially associated with pure spirit and female attributes constructed as merely material, bereft of spirit. As Johann Jacob Bachofen has stated, "The triumph of paternity brings with it the liberation of the spirit from the manifestations of nature, a sublimation of human existence over the laws of material life. Maternity pertains to the physical side of man, the only thing he shares with animals; the paternal spiritual principle belongs to him alone. Triumphant paternity partakes of the heavenly light, while child-bearing motherhood is bound up with the earth that bears all things."[1]

Central to the patriarchal assumption of men's superiority over women is the social construct of passivity/materiality as female and animal, and activity/spirituality as male and distinctly human. This is reflected in dualisms like mind/body, with the mind being nonmaterial, male, and active, and the body physical, female, and passive. It is also reflected in the dualism of culture/nature, with the assumption that men alone have access to culture while women are bound up with the earth that bears all things.[2] What these artificial dichotomies obscure is that activity, not passivity, is nature's forte.

The new biotechnologies reproduce the old patriarchal divisions of activity/passivity, culture/nature. These dichotomies are then used as instruments of capitalist patriarchy to colonize the regeneration of plants and human beings. Only by decolonizing regeneration can the activity and creativity of women and nature in a nonpatriarchal mold be reclaimed.

Organisms, the New Colonies

The land, the forests, the rivers, the oceans, and the atmosphere have all been colonized, eroded, and polluted. Capital now has to look for new colonies to invade and exploit for its further accumulation—the interior spaces of the bodies of women, plants, and animals.

The invasion and takeover of land as colonies was made possible through the technology of the gunboat; the invasion and takeover of the life of organisms as the new colonies is being made possible through the technology of genetic engineering.

Biotechnology, as the handmaiden of capital in the postindustrial era, makes it possible to colonize and control that which is autonomous, free, and self-regenerative. Through reductionist science, capital goes where it has never been before. The fragmentation of reductionism opens up areas for exploitation and invasion. Technological development under capitalist patriarchy proceeds steadily from what it has already transformed and used up, driven by its predatory appetite, toward that which has still not been consumed. It is in this sense that the seed and women's bodies as sites of regenerative power are, in the eyes of capitalist patriarchy, among the last colonies.[3]

While ancient patriarchy used the symbol of the active seed and the passive earth, capitalist patriarchy, through the new biotechnologies, reconstitutes the seed as passive, and locates activity and creativity in the engineering mind. Five hundred years ago, when land began to be colonized, the reconstitution of the earth from a living system into mere matter went hand in hand with the devaluation of the contributions of non-European cultures and nature. Now, the reconstitution of the seed from a regenerative source of life into valueless raw material goes hand in hand with the devaluation of those who regenerate life through the seed—that is, the farmers and peasants of the third world.

From Terra Mater to Terra Nullius

All sustainable cultures, in their diversity, have viewed the earth as terra mater. The patriarchal construct of the passivity of the earth and the consequent creation of the colonial category of land as terra nullius served two purposes: it denied the existence and prior rights of original inhabitants, and it negated the regenerative capacity and life processes of the earth.[4] The decimation of indigenous peoples everywhere was justified morally on the grounds that they were not really human; they were part of the fauna. As John Pilger has observed, the *Encyclopaedia*

Britannica appeared to be in no doubt about this in the context of Australia: "Man in Australia is an animal of prey. More ferocious than the lynx, the leopard, or the hyena, he devours his own people."[5] In an Australian textbook, *Triumph in the Tropics,* Australian aborigines were equated with their half-wild dogs.[6] Being animals, the original Australians and Americans, the Africans and Asians, possessed no rights as human beings. Their lands could be usurped as terra nullius—lands empty of people, vacant, wasted, and unused. The morality of the missions justified the military takeover of resources all over the world to serve imperial markets. European men were thus able to describe their invasions as discoveries, their piracy and theft as trade, and their extermination and enslavement as a civilizing mission.

Scientific missions colluded with religious missions to deny rights to nature. The rise of mechanical philosophy with the emergence of the scientific revolution was based on the destruction of concepts of a self-regenerative, self-organizing nature that sustained all life. For Francis Bacon, who is called the father of modern science, nature was no longer a mother but rather, a female to be conquered by an aggressive masculine mind. As Carolyn Merchant points out, this transformation of nature from a living, nurturing mother to inert, dead, and manipulable matter was eminently suited to the exploitation imperative of growing capitalism. The nurturing earth image acted as a cultural constraint on the exploitation of nature. "One does not readily slay a mother, dig her entrails, or mutilate her body," writes Merchant. But the images of mastery and domination created by the Baconian program and the scientific revolution removed all restraint, and functioned as cultural sanctions for the denudation of nature.

The removal of animistic, organic assumptions about the cosmos constituted the death of nature—the most far-reaching effect of the scientific revolution. Because nature was now viewed as a system of dead, inert particles moved by external rather than inherent forces, the mechanical framework itself could legitimize the manipulation of nature. Moreover, as a conceptual framework, the mechanical order was associated with a framework of values based on power, fully compatible with the directions taken by commercial capitalism.[7]

The construct of the inert earth was given a new and sinister significance as development denied the earth's productive capacity and created systems of agriculture that could not regenerate or sustain themselves.

Sustainable agriculture is based on the recycling of soil nutrients. This involves returning to the soil part of the nutrients that come from it and support plant growth. The maintenance of the nutrient cycle, and through it the fertility of the soil, is based on an inviolable law of return that recognizes the earth as the source of fertility. The Green Revolution paradigm of agriculture substituted the regenerative nutrient cycle with linear flows of purchased inputs of chemical fertilizers from factories and marketed outputs of agricultural commodities. Fertility was no longer the property of soil, but of chemicals. The Green Revolution was essentially based on miracle seeds that needed chemical fertilizers and did not produce plant outputs for returning to the soil.[8] The earth was again viewed as an empty vessel, this time for holding intensive inputs of irrigated water and chemical fertilizers. The activity lay in the miracle seeds, which transcended nature's fertility cycles.

Ecologically, however, the earth and soil were not empty, and the growth of Green Revolution varieties did not take place only with the seed fertilizer packet. The creation of soil diseases and micronutrient deficiencies was an indication of the invisible demands the new varieties were making on the fertility of the soil; desertification indicated the broken cycles of soil fertility caused by an agriculture that produced only for the market. The increase in production of grain for marketing was achieved in the Green Revolution strategy by reducing the biomass for internal use on the farm. The reduction of output for straw production was probably not considered a serious cost since chemical fertilizers were thought to be a total substitute for organic manure. Yet, as experience has shown, the fertility of soils cannot be reduced to nitrogen, phosphorus, and potassium in factories, and agricultural productivity necessarily includes returning to the soil part of the biological products that the soil yields. The seed and the earth mutually create conditions for each other's regeneration and renewal. Technologies cannot provide a substitute for nature and cannot work outside of nature's ecological processes without destroying the very basis of production, nor can markets provide the only measure of output and yield.

Seeds from the Lab

While the Green Revolution was based on the assumption that the earth is inert, the biotechnology revolution robs the seed of its fertility and self-regenerative capabilities, colonizing it in two major ways: through technical means and through property rights.

Processes like hybridization are the technological means that stop seed from reproducing itself. This provides capital with an eminently effective way of circumventing natural constraints on the commodification of the seed. Hybrid varieties do not produce true-to-type seed, and farmers must return to the breeder each year for new seed stock.

To use Jack Kloppenburg's description of the seed: it is both a means of production and a product.[9] Whether they are tribespeople engaged in shifting cultivation or peasants practicing settled agriculture, in planting each year's crop, farmers also reproduce the necessary element of their means of production. The seed thus presents capital with a simple biological obstacle: given the appropriate conditions, it reproduces itself and multiplies. Modern plant breeding has primarily been an attempt to remove this biological obstacle, and the new biotechnologies are the latest tools for transforming what is simultaneously a means of production and a product into mere raw material.

The hybridization of seed was an invasion into the seed itself. As Kloppenburg has stated, it broke the unity of the seed as food grain and as a means of production. In doing so, it opened up the space for capital accumulation that private industry needed in order to control plant breeding and commercial seed production. And it became the source of ecological disruption by transforming a self-regenerative process into a broken linear flow of supply of living seed as raw material and a reverse flow of seed commodities as products. The decoupling of seed from grain also changes the status of seed.

The commodified seed is ecologically incomplete and ruptured at two levels: first, it does not reproduce itself, while by definition, seed is a regenerative resource. Genetic resources are thus, through technology, transformed from a renewable into a nonrenewable resource. Second, it does not produce by itself; it needs the help of other purchased inputs. And as the seed and chemical companies merge, the dependence on inputs will increase. Whether a chemical is added externally or internally, it remains an external input in the ecological cycle of the repro-duction of seed. It is this shift from ecological processes of production through regeneration to technological processes of nonregenerative production that underlies the dispossession of farmers and the dras-tic reduction of biological diversity in agriculture. It is at the root of the creation of poverty and of nonsustainability in agriculture.

Where technological means fail to prevent farmers from repro-ducing their own seed, legal regulations in the forms of intellectual property rights and patents are brought in. Patents are central to the colonization of plant regeneration and, like land titles, are based on the assumption of ownership and property. As the vice president of Genen-tech has stated, "When you have a chance to write a clean slate, you can make some very basic claims, because the standard you are com-pared to is the state of prior art, and in biotechnology there just is not much."[10] Ownership and property claims are made on living resources, but prior custody and use of those resources by farmers is not the mea-sure against which the patent is set. Rather, it is the intervention of tech-nology that determines the claim to their exclusive use. The possession of this technology, then, becomes the reason for ownership by corpora-tions, and for the simultaneous dispossession and disenfranchisement of farmers.

As with the transformation of terra mater to terra nullius, the new biotechnologies rob farmers' seeds of life and value by the very process that makes corporate seeds the basis of wealth creation. Indigenous vari-eties, called landraces, evolved through both natural and human selec-tion; produced and used by third world farmers worldwide, they are primitive cultivars. Those varieties created by modern plant breeders in international research centers or by transnational seed corporations are called advanced or elite. Trevor Williams, the former executive secre-tary of the International Board for Plant Genetic Resources, has argued that it is not the original material that produces cash returns. At a 1983 forum on plant breeding, he stated that raw germ plasm only becomes valuable after considerable investment of time and money.[11] According to this calculation, peasants' time is considered valueless and available for free. Once again, all prior processes of creation are being denied and devalued by defining them as nature. Thus, plant breeding by farm-ers is not breeding; real breeding is seen to begin when this "primitive germ plasm" is mixed or crossed with inbred lines in international labs by international scientists. That is, innovation occurs only through the long, laborious, expensive, and always risky process of backcrossing and other means required to first make genetic sense out of the chaos created by the foreign germ plasm, and eventually to make dollars and cents from a marketable product.[12]

But the landraces that farmers have developed are not genetically chaotic. Nor do they lack innovation. They consist of improved and selected material, embodying the experience, inventiveness, and hard work of farmers, past and present; the evolutionary material processes they have undergone serve ecological and social needs. These needs are now being undermined by the monopolizing tendency of corporations. Placing the contributions of corporate scientists over and above the intellectual contributions made by third world farmers over ten thousand years—contributions to conservation, breeding, domestication, and development of plant and animal genetic resources—is based on rank social discrimination.

IPRs versus Farmers' and Plant Breeders' Rights

As Pat Mooney has argued, "The perception that intellectual property is only recognizable when produced in laboratories by men in lab coats is fundamentally a racist view of scientific development."[13] Indeed, the total genetic change achieved by farmers over millennia has been far greater than that achieved during the last one hundred to two hundred years of more systematic science-based efforts. The limits of the market system in assigning value can hardly be a reason for denying value to farmers' seeds and nature's seeds. It indicates the deficiencies in the logic of the market rather than the status of the seed or of farmers' intelligence.

The denial of prior rights and creativity is essential for owning life. A brief book prepared by the biotechnology industry states: "Patent laws would in effect have drawn an imaginary line around your processes and products. If anyone steps over that line to use, make or sell your inventions or even if someone steps over that line in using, making or selling his own products, you could sue for patent protection."[14] Jack Doyle has appropriately remarked that patents are less concerned with innovation than with territory, and can act as instruments of territorial takeover by claiming exclusive access to creativity and innovation, thereby monopolizing rights to ownership.[15] The farmers, who are the guardians of the germ plasm, have to be dispossessed to allow the new colonization to happen.

As with the colonization of land, the colonization of life processes will have a serious impact on third world agriculture. First, it will undermine the cultural and ethical fabric of agriculturally based societies. For instance, with the introduction of patents, seeds—which have hitherto been treated as gifts and exchanged freely between farmers—will become patented commodities. Hans Leenders, former secretary general of the International Association of Plant Breeders for the Protection of Plant Varieties, has proposed the abolition of the farmer's right to save seed. He says: "Even though it has been a tradition in most countries that a farmer can save seed from his own crop, it is under the changing circumstances not equitable that farmers can use this seed and grow a commercial crop out of it without payment of a royalty; the seed industry will have to fight hard for a better kind of protection."[16]

Although genetic engineering and biotechnology only relocate existing genes rather than create new ones, the ability to relocate and separate is translated into the power and right to own. The power to own a part is then translated into control of the entire organism.

Additionally, the corporate demand for the conversion of a common heritage into a commodity, and for profits generated through this transformation to be treated as property rights, will have serious political and economic implications for third-world farmers. They will now be forced into a three-level relationship with the corporations demanding a monopoly on life-forms and life processes through patents. First, farmers are suppliers of germ plasm to transnational corporations; second, they become competitors in terms of innovation and rights to genetic resources; and third, they are consumers of the technological and industrial products of these corporations. In other words, patent protection transforms farmers into suppliers of free raw material, displaces them as competitors, and makes them totally dependent on industrial supplies for vital inputs such as seed. The frantic cry for patent protection in agriculture is really a ruse for control of biological resources in agriculture. It is argued that patent protection is essential for innovation, yet it is essential only for corporate profits. After all, farmers have been making innovations for centuries, as have public institutions for decades, without property rights or patent protection.

Further, unlike plant breeders' rights (PBRs), the new utility patents are very broad based, allowing monopoly rights over individual genes and even characteristics. PBRs do not entail ownership of the germ plasm in the seeds; they only grant a monopoly right over the selling and marketing of a specific variety. Patents, on the other hand, allow for multiple claims that may cover not only whole plants but plant parts and processes as well. So, according to attorney Anthony Diepenbrock: "You could file for protection of a few varieties of crops, their macroparts (flowers, fruits, seeds and so on), their microparts (cells, genes, plasmids and the like) and whatever novel processes you develop to work these parts, all using one multiple claim."[17]

Patent protection implies the exclusion of farmers' rights over resources having these genes and characteristics, undermining the very foundation of agriculture. For example, a patent has been granted in the United States to a biotechnology company, Sungene, for a sunflower variety with very high oleic acid content. The claim allowed was for the characteristic (i.e., high oleic acid) and not just for the genes producing the characteristic. Sungene has notified sunflower breeders that the development of any variety high in oleic acid will be considered an infringement of its patent.

The landmark event for the patenting of plants was the 1985 judgment in the United States, now famous as ex parte Hibberd, in which molecular genetics scientist Kenneth Hibberd and his coinventors were granted patents on the tissue culture, seed, and whole plant of a corn line selected from tissue culture.[18] The Hibberd application included over 260 separate claims, which gave the molecular genetics scientists the right to exclude others from use of all 260 aspects. While the Hibberd case apparently provides a new legal context for corporate competition, the most profound impact will be felt in the competition between farmers and the seed industry.

As Kloppenburg has indicated, with the Hibberd judgment, a juridical framework is now in place to allow the seed industry to realize one of its longest-held and most cherished goals: to force all farmers to buy seed every year instead of obtaining it through reproduction. Industrial patents allow others to use a product but deny them the right to make it. Since seed makes itself, a strong utility patent for seed implies that a farmer purchasing patented seed would have the right to use (to grow) the seed, but not to make it (to save and replant). If the Dunkel Draft of the GATT is implemented, the farmer who saves and replants the seed of a patented or protected plant variety will be violating the law.

Through intellectual property rights, an attempt is made to take away what belongs to nature, to farmers, and to women, and to term this invasion improvement and progress. Violence and plunder as instruments of wealth creation are essential to the colonization of nature and of our bodies through the new technologies. Those who are exploited become the criminals; those who exploit require protection. The North must be protected from the South so that it can continue its uninterrupted theft of the third world's genetic diversity. The seed wars, trade wars, patent protection, and intellectual property rights at the GATT are claims to ownership through separation and fragmentation. If the regime of rights being demanded by the United States is implemented, the transfer of funds from poor to rich countries will exacerbate the third world crisis ten times over.[19]

The United States has accused the third world of piracy. The estimates for royalties lost are $202 million per year for agricultural chemicals and $2.5 billion annually for pharmaceuticals.[20] In a 1986 U.S. Department of Commerce survey, U.S. companies claimed they lost $23.8 billion yearly due to inadequate or ineffective protection of intellectual property. Yet, as the team at the Rural Advancement Foundation International in Canada has shown, if the contributions of third world peasants and tribespeople are taken into account, the roles are dramatically reversed: the United States would owe third-world countries $302 million in agriculture royalties and $5.1 billion for pharmaceuticals. In other words, in these two biological industry sectors alone, the United States should owe $2.7 billion to the third world.[21] It is to prevent these debts from being taken into account that it becomes essential to set up the creation boundary through the regulation of intellectual property rights; without it, the colonization of the regenerative processes of life renewal is impossible. Yet if this, too, is allowed to happen in the name of patent protection, innovation, and progress, life itself will have been colonized.

There are, at present, two trends reflecting different views as to how native seeds, indigenous knowledge, and farmers' rights should be treated. On the one hand, there are initiatives across the world that recognize the inherent value of seeds and biodiversity, acknowledge the contributions of farmers to agricultural innovation and seed conservation, and see patents as a threat both to genetic diversity and to farmers. At the global level, the most significant platforms to have made the issue of farmers' rights visible are the Food and Agriculture Organization (FAO) Commission of Plant Genetic Resources[22] and the Keystone Dialogue.[23] At the local level, communities all over Asia, Africa, and Latin America are taking steps to save and regenerate their native seeds. To mention one example, we have set up a network in India called Navdanya with the goal of native seed conservation.

Despite these initiatives, however, the dominant trend continues to be toward the displacement of local plant diversity and its substitution by patented varieties. At the same time, international agencies under pressure from seed corporations are pushing for regimes of intellectual property rights that deny farmers their intellect and their rights. The March 1991 revision of the International Convention for the Protection of New Varieties of Plants, for example, allows countries to remove the farmers' exemption—the right to save and replant seed—at their discretion.[24]

In another development leading to the privatization of genetic resources, the Consultative Group on International Agricultural Research made a policy statement on May 22, 1992, allowing for the privatization and patenting of genetic resources held in international gene banks.[25] The strongest pressure for patents is coming from the GATT, especially in relation to the agreement on TRIPs and agriculture.[26]

Engineering Humans

Just as technology changes seed from a living, renewable resource into mere raw material, it devalues women in a similar way. For instance, reproduction has been linked to the mechanization of the female body, in which a set of fragmented, fetishized, and replaceable parts is managed by professional medical experts. While this tendency is most advanced in the United States, it is also spreading to the third world.

The mechanization of childbirth is evident in the increased use of Caesarean sections. Significantly, this method, which requires the most management by the doctor and the least labor by the woman, is seen as providing the best product. But Caesarean sections are a surgical procedure, and the chances of complications are two to four times greater than during normal vaginal delivery. They were introduced as a means of delivering babies at risk, but when they are done routinely, they can pose an unnecessary threat to health and even life. Close to one in every four Americans is now born by Caesarean section.[27] Brazil has one of the highest proportions of Caesarean section deliveries in the world; a nationwide study of patients enrolled in the social security system showed an increase in the proportion of Caesarean sections from 15 percent in 1974 to 31 percent in 1980. In urban areas, such as the city of Sao Paulo, rates as high as 75 percent have been observed.

As with plant regeneration, where agriculture has moved from the Green Revolution technologies to biotechnology, a parallel shift is also taking place with regard to human reproduction. With the introduction of new reproductive technologies, the relocation of knowledge and skills from the mother to the doctor, from women to men, will be accentuated. Peter Singer and Deane Wells, in *The Reproductive Revolution*, have suggested that the production of sperm is worth a great deal more than the production of eggs. They conclude that sperm vending places a greater strain on the man than egg donation does on the woman, in spite of the chemical and mechanical invasion of her body.[28]

While currently, in vitro fertilization and other technologies are offered for abnormal cases of infertility, the boundary between nature and nonnature is fluid, and normality has a tendency to be redefined as abnormality when technologies created for abnormal cases become more widely used. When pregnancy was first transformed into a medical disease, professional management was limited to abnormal cases, while normal cases continued to be looked after by the original professional, the midwife. While 70 percent of childbirths in the United Kingdom were thought normal enough to be delivered at home in the 1930s, by the 1950s, the same percentage were identified as abnormal enough to be delivered in the hospital!

The new reproductive technologies have provided contemporary scientific rhetoric for the reassertion of an enduring set of deeply patriarchal beliefs. The idea of women as vessels, and the fetus as created by the father's seed and owned by patriarchal right, leads logically to the breaking of organic links between the mother and the fetus.

Medical specialists, falsely believing that they produce and create babies, force their knowledge on knowing mothers. They treat their own knowledge as infallible, and women's knowledge as wild hysteria. And through their fragmented and invasive knowledge, they create a maternal-fetal conflict in which life is seen only in the fetus, and the mother is reduced to a potential criminal threatening her baby's life.

The false construction of a maternal-fetal conflict, which was the basis of the patriarchal takeover of childbirth by male medical practitioners from women and midwives, was adopted by feminists as the basis of women's "choice" a century later. The "pro-choice" and "pro-life" movements are, thus, both based on a patriarchal construction of women and reproduction.

The medical construction of life through technology is often inconsistent with the living experience of women as thinking and knowing human beings. When such conflicts arise, patriarchal science and law have worked hand in hand to establish the control by professional men over women's lives, as demonstrated by recent work on surrogacy and the new reproductive technologies. Women's rights, linked with their regenerative capacities, have been replaced by those of doctors as producers and rich, infertile couples as consumers.

The woman whose body is being exploited as a machine is not seen as the one who needs protection from doctors and rich couples. Instead, the consumer, the adoptive male parent, needs protection from the biological mother, who has been reduced to a surrogate uterus. This is exemplified in the famous 1986 Baby M. case, in which Mary Beth agreed to lend her uterus, but after experiencing what having a baby meant, wanted to return the money and keep the child. A New Jersey judge ruled that a man's contract with a woman concerning his sperm is sacred, and that pregnancy and childbirth are not. Commenting on this notion of justice, in her book *Sacred Bond*, Phyllis Chesler says, "It's as if these experts were 19th century missionaries and Mary Beth a particularly stubborn native who refused to convert to civilization, and what's more, refused to let them plunder her natural resources without a fight."[29]

The role of man as creator has also been taken to absurd lengths in an application submitted for a patent for the characterization of the gene sequence coding for human relaxin, a hormone that is synthesized and stored in female ovaries and helps in dilation, thus facilitating the birth process. A naturally occurring substance in women's bodies is being treated as the invention of three male scientists, Peter John Hud, Hugh David Nill, and Geoffrey William Tregear.[30] Ownership is thus acquired through invasive and fragmenting technology, and it is this link between fragmenting technology and control and ownership of resources and people that forms the basis of the patriarchal project of knowledge as power over others.

Such a project is based on the acceptance of three separations: the separation of mind and body; the gendered separation of male activity as intellectual and female activity as biological; and the separation of the knower and the known. These separations allow for the political construction of a creation boundary that divides the thinking, active male from the unthinking, passive female, and from nature.

Biotechnology is today's dominant cultural instrument for carving out the boundary between nature and culture through intellectual property rights, and for defining women's and farmers' knowledge and work as nature. These patriarchal constructs are projected as natural, although there is nothing natural about them. As Claudia Von Werlhof has pointed out, from the dominant standpoint, nature is everything that should be available free or as cheaply as possible. This includes the products of social labor. The labor of women and third world farmers is said to be nonlabor, mere biology, a natural resource; their products are thus akin to natural deposits.[31]

The Production and Creation Boundaries

The transformation of value into disvalue, labor into nonlabor, knowledge into nonknowledge, is achieved by two very powerful constructs: the production boundary and the creation boundary.

The production boundary is a political construct that excludes regenerative, renewable production cycles from the domain of production. National accounting systems, which are used for calculating growth through the gross national product, are based on the assumption that if producers consume what they produce, they do not, in fact, produce at all because they fall outside of the production boundary.[32] All women who produce for their families, children, and nature are thus treated as nonproductive, as economically inactive. Discussions at the United Nations Conference on Environment and Development on issues of biodiversity have also referred to production for one's own consumption as a market failure.[33] Self-sufficiency in the economic domain is, therefore, seen as economic deficiency when economies are confined to the marketplace. The devaluation of women's work, and of work done in subsistence economies in the third world, is the natural outcome of a production boundary constructed by capitalist patriarchy.

The creation boundary does to knowledge what the production boundary does to work: it excludes the creative contributions of women as well as third world peasants and tribespeople and views them as being engaged in unthinking, repetitive biological processes. The separation of production from reproduction and the characterization of the former as economic and the latter as biological are some of the underlying assumptions that are treated as natural even though they have been socially and politically constructed.

This patriarchal shift in the creation boundary is misplaced for many reasons. First, the assumption that male activity is true creation because it takes place ex nihilo is ecologically false. No technological artifact or industrial commodity is formed out of nothing; no industrial process takes place where nothing was before. Nature and its creativity as well as people's social labor are consumed at every level of industrial production as raw material or energy. The biotech seed that is treated as a creation to be protected by patents could not exist without the farmer's seed. The assumption that only industrial production is truly creative because it produces from nothing hides the ecological destruction that goes with it. The patriarchal creation boundary allows ecological destruction to be perceived as creation, and ecological regeneration as underlying the breakdown of ecological cycles and the crisis of sustainability. To sustain life means, above all, to regenerate life; but according to the patriarchal view, to regenerate is not to create, it is merely to repeat.

Such a definition of creativity is also false because it fails to see that women's and subsistence producers' work go into childrearing and cultivation, both of which conserve regenerative capacity.

The assumption of creation as the reduction of novelty is also false;

regeneration is not merely repetition. It involves diversity, while engineering produces uniformity. Regeneration, in fact, is how diversity is produced and renewed. While no industrial process takes place out of nothing, the creation myth of patriarchy is particularly unfounded in the case of biotechnologies, where life-forms are the raw material for industrial production.

Rebuilding Connections

The source of patriarchal power over women and nature lies in separation and fragmentation. Nature is separated from and subjugated to culture; mind is separated from and elevated above matter; female is separated from male, and identified with nature and matter. The domination over women and nature is one outcome; the disruption of cycles of regeneration is another. Disease and ecological destruction arise from this interruption of the cycles of renewal of life and health. The crises of health and ecology suggest that the assumption of man's ability to totally engineer the world, including seeds and women's bodies, is in question. Nature is not the essentialized, passive construct that patriarchy assumes it to be. Ecology forces us to recognize the disharmonies and harmonies in our interactions with nature. Understanding and sensing connections and relationships is the ecological imperative.

The main contribution of the ecology movement has been the awareness that there is no separation between mind and body, human and nature. Nature consists of the relationships and connections that provide the very conditions for our life and health. This politics of connection and regeneration provides an alternative to the politics of separation and fragmentation that is causing ecological breakdown. It is a politics of solidarity with nature. This implies a radical transformation of nature and culture in such a manner that they are mutually permeating, not separate and oppositional. By starting a partnership with nature in the politics of regeneration, women are simultaneously reclaiming their own and nature's activity and creativity. There is nothing essentialist about this politics because it is, in fact, based on denying the patriarchal definition of passivity as the essence of women and nature. There is nothing absolutist about it because the natural is constructed through diverse relationships in diverse settings. Natural agriculture and natural childbirth involve human creativity and sensitivity of the highest order, a creativity and knowledge emerging from partnership and participation, not separation. The politics of partnership with nature, as it is being shaped in the everyday lives of women and communities, is a politics of rebuilding connections, and of regeneration through dynamism and diversity.

Notes

1. Johann Jacob Bachofen, quoted in Marta Weigle, *Creation and Procreation* (Philadelphia: University of Pennsylvania Press, 1989).

2. Ibid.

3. Claudia Von Werlhof, "Women and Nature in Capitalism," in *Women: The Last Colony,* ed. Maria Mies (London: Zed Books, 1989).

4. John Pilger, *A Secret Country* (London: Vintage, 1989).

5. Ibid.

6. Ibid.

7. Carolyn Merchant, *The Death of Nature: Women, Ecology and the Scientific Revolution* (New York: Harper and Row, 1980).

8. Vandana Shiva, *The Violence of the Green Revolution* (Penang: Third World Network, 1991).

9. Jack Kloppenburg, *First the Seed* (Cambridge: Cambridge University Press, 1988).

10. Quoted in Jack Doyle, *Altered Harvest* (New York: Viking, 1985), 310.

11. Kloppenburg, *First the Seed,* 85.

12. Stephen Witt, "Biotechnology and Genetic Diversity," California Agricultural Lands Project, San Francisco, 1985.

13. Pat Mooney, "From Cabbages to Kings," *Development Dialogue* (1988): 1–2, and *Proceedings of the Conference on Patenting of Life Forms* (Brussels: ICDA, 1989).

14. Witt, "Biotechnology and Genetic Diversity."

15. Doyle, *Altered Harvest.*

16. Hans Leenders, "Reflections on 25 Years of Service to the International Seed Trade Federation," *Seedsmen's Digest* 37, no. 5 (1986): 89.

17. Quoted in Kloppenburg, *First the Seed,* 266.

18. Ibid.

19. Rural Advancement Foundation International, *Biodiversity, UNICED and GATT* (Ottawa, 1991).

20. Ibid.

21. Ibid.

22. Food and Agriculture Organization (FAO), *International Undertaking on Plant Genetic Resources,* DOC C83/II REP/4 and 5 (Rome, 1983).

23. Keystone International Dialogue on Plant Genetic Resources, Final Con-

sensus Report of Third Plenary Session, Keystone Center, Colorado, May 31–June 4, 1991.

24. Genetic Resources Action International (GRAIN), "Disclosures: UPOV Sells Out," Barcelona, Spain, December 2, 1990.

25. Vandana Shiva, "Biodiversity, Biotechnology and Bush," in *Third World Network Earth Summit Briefings* (Penang: Third World Network, 1992).

26. Vandana Shiva, "GATT and Agriculture," *Bombay Observer,* 1992.

27. Neil Postman, *Technology: The Surrender of Culture to Technology* (New York: Knopf, 1992).

28. Peter Singer and Deane Wells, *The Reproductive Revolution: New Ways of Making Babies* (Oxford: Oxford University Press, 1984).

29. Phyllis Chesler, *Sacred Bond: Motherhood under Siege* (London: Virago, 1988).

30. European Patent Office, application no. 833075534.

31. Von Werlhof, "Women and Nature in Capitalism."

32. Marilyn Waring, *If Women Counted* (New York: Harper and Row, 1988).

33. United Nations Conference on Environment and Development, "Agenda 21," adopted by the plenary on June 14, 1992, published by the UNCED Secretariat, Conches, Switzerland.

9

Seeds of Suicide

The Ecological and Human Costs
of the Globalization of Agriculture

This study takes stock of the impact of a decade of trade liberalization on the lives and livelihood of farmers. Across India farmers are taking the desperate step of ending their lives because of new pressures building up on them as a result of globalization and the corporate takeover of seed supply, leading to the spread of capital-intensive agriculture and the propagation of nonsustainable agriculture practices. The lure of huge profits linked with clever advertising strategies evolved by the seed and chemical industries and easy credit for the purchase of costly inputs is forcing farmers into a chemical treadmill and a debt trap. The reality of globalization is different from the corporate propaganda and from the promises of trade liberalization and agriculture offered by the World Bank, the WTO, and experts and economists sitting in our various ministries. The impacts of trade liberalization and globalization have been felt in each and every state of India, with the states of Andhra Pradesh, Karnataka, Maharashtra, and Punjab bearing the maximum burden in terms of the high social and ecological costs. Farmers are paying for globalization by being forced to sacrifice their lives and livelihoods. In what follows we present scenarios from these states on the status of farmers' suicides; since December 1997, when such suicides first reached epidemic proportions, the Research Foundation for Science, Technology, and Ecology has been monitoring these incidents and analyzing their causes.

The epidemic of farmers' suicide is a barometer of the stress Indian agriculture and Indian farmers have been put under by the globalization of agriculture. Growing indebtedness and increasing crop failure are the main reasons farmers have committed suicide across the length and breath of rural India. These are also the inevitable outcomes of the corporate model of industrial agriculture being introduced in India through globalization. Agriculture driven by multinational corporations is capital intensive and creates heavy debt for the purchase of costly inputs such as seeds and agrochemicals. It is also ecologically vulnerable since it is based on monocultures of introduced varieties and on nonsustainable practices of chemically intensive farming.

The high social and ecological costs of the globalization of nonsustainable agriculture are not restricted to the cotton-growing areas of the states under consideration here; they have been experienced in all regions where commercially grown and chemically farmed crops are raised. While the benefits of globalization go to the seed and chemical corporations through expanding markets, the cost and risks are exclusively borne by the small farmers and landless peasants.

The two most significant ways through which the risks of crop failures have been increased by globalization are the introduction of ecologically vulnerable hybrid seeds and the increased dependence on agrochemical input such as pesticides, necessary with pest-prone hybrids.

The privatization of the seed sector under trade liberalization has led to a shift in cropping patterns from polyculture to monoculture and a shift from open-pollinated varieties to hybrids. In the district of Warangal in Andhra Pradesh, this shift has been very rapid, converting Warangal from a mixed-farming system based on millets, pulses, and oilseeds to a monoculture of hybrid cotton.

The problem of pests is a problem created by the erosion of diversity in crops and cropping patterns and the introduction of commercial hybrid seeds. The most sustainable solution for pest control is rejuvenating biodiversity in agriculture. Nonsustainable pest control strategies offer chemical or genetic fixes while reducing diversity, which is the biggest insurance against pest damage.

As the cotton disaster described below shows, the globalization of agriculture is threatening both the environment and the survival of farmers. Biodiversity is being destroyed, the use of agrochemicals is increasing, ecological vulnerability is increasing, and farmer debts are skyrocketing, leading to suicides in extreme cases.

The Andhra Pradesh Scenario of Farmers' Suicides

From Mixed Farming to Monocultures: The Lure of "White Gold"

More than sixteen thousand farmers have committed suicide in Andhra Pradesh alone from 1995 to 1997.[1] Taking into consideration the large number of suicides during 1998 and 1999, it is possible that by early 2001 the number will reach twenty thousand.

Cotton cultivation has been taken up in areas that were not traditionally used for cotton growing. One such region is Warangal district in Andhra Pradesh, which has switched from predominantly food crops to cotton, which is a relatively new crop brought in under trade liberalization. The area given over to cotton in this region grew over threefold within a decade. In Warangal, over three decades (between the 1960s and the 1980s), the total acreage devoted to cotton was negligible. According to the available data, in 1986–1987 the total area under cotton cultivation was 32,792 hectares (or 81,980 acres), which increased to 100,646 hectares (or 251,615 acres) in 1996–1997, nearly three times as much. The cotton cultivation has basically replaced the jawar crop. The area under jawar in 1986–1987 was 77,884 hectares, which went down to 27,306 hectares in 1996–1997. The acreage under the traditional paddy has also shrunk. The land under bajra (millet) has been drastically reduced in the last ten years, from 11,289 hectares in 1986–1987 to just 400 hectares in 1996–1997.

The acreage under cotton increased because the farmers in Warangal were getting a good return on cotton. But in the 1997–1998 season there was heavy damage to the cotton crop. There were several reasons for the crop failure, but the most important were bad weather and a severe pest attack. There was drought in June–July, the main sowing season for cotton. Because of the drought only 15 percent paddy could be planted. In October–November the rain came during the cotton boll-bursting season. The untimely rain also affected the paddy because it was in the maturity period. The cloudy weather, untimely rain, and lack of winter in November–December led to the emergence of pests.

In 1997 the pests first emerged in the chili fields, and the weather helped them to multiply. The pests attacked all the crops in the fields— chili, cotton, red gram, and so on—the yield was thus heavily reduced.

G. Mahendar of Mulkaligud village in Warangal district bought Excel cottonseed last year after being lured by the company's advertising propaganda. The company dealers took two jeep loads of farmers to the trial fields of Excel cotton every day and informed them that the variety yielded eighteen quintals per acre. Farmers in Mulkaligud and neighboring villages planted thirty-five thousand acres of land with the variety. The crop did not perform well; the plant shed the bolls it developed.

The farmers complained to the dealer in their area and demanded compensation. Many company officials visited the farmers' households and conducted elaborate surveys. However, so far no compensation has been paid to the farmers. Instead the dealer threatened to close the shop in their area and open a new shop elsewhere.

Since several sprays of chemicals had already been made by that time, they had no effect on the pests. The more the chemicals failed, the more they were used. The panic created by the pests led to heavy dosages of pesticides sprayed at frequent intervals in the cotton fields.

Analysis of the cotton failure has been focused on the excessive use of pesticides or of spurious pesticides. However, pesticide use is intimately linked to hybrid seeds. Pesticides become necessary when crop varieties and cropping patterns are vulnerable to pest attacks. Hybrid seeds offer a promise of higher yields, but they also have higher risks of crop failure since they are more prone to pest and disease attack, as illustrated by the Andhra Pradesh experience. Monocultures further increase the vulnerability to pest attacks since the same crop of the same variety planted over large areas year after year encourages pest buildups.

Privatization and the Spread of Monocultures

No native variety of cotton is found in Warangal—all varieties of cottonseeds used in Warangal are hybrid seeds sold by private companies providing high-yielding varieties. It takes at least six to seven years of trials and verifications under the supervision of government authorities for any company to launch certified seeds. However, to avoid such delays in the introduction of seeds to the market, seed companies sell them as "truthful" seeds: seeds sold on the basis of farmers having confidence in the company's claims about them. There is no regulation to prevent the sale of "truthful" seeds.

In the 1970s cotton cultivation in Warangal was dependent upon the varieties developed by the public sector seed supply. During that time the most popular variety was Hybrid-4, a short staple cotton variety. Besides Hybrid-4 (H-4), the other varieties used during the 1970s and 1980s were MCU-5 (developed by Coimbatore Research Station); L. K. varieties (which were resistant to white fly and jassids); Varalakshmi (developed by Cotton Research Station, Nandyal); and JKHY-1 (an HYV developed by Jawaharlal Nehru Krishi Vidhyalaya, MP), among others. All these varieties were government varieties that were cultivated in the Telengana region.

However, during the 1980s a handful of private companies participated in cotton research and evolved a number of hybrid cotton varieties. These included Maharashtra Hybrid Seeds Company, Jalna (Mahyco); Mahindra Seeds Company, Jalna; and Nath Seeds Company, Aurangabad. These companies captured the entire market for cottonseed production and distribution.

The most popular variety of cotton in Warangal, based on yields during 1995–1997, was RCH-2, a long-duration "truthful" hybrid variety produced by Rasi Seeds Company and marketed by J. K. Company, Secundrabad. Other varieties of cottonseeds grown by farmers and the acreage under each variety in Warangal during 1996–1997 are shown in table 9.2. In Adilabad the most popular variety during this period was the short-duration L. K. variety. The MCU varieties were popular in Khammam district. The choice of variety for a particular region depends upon its soil condition, water availability, and the inclination of farmers. As a result of aggressive marketing by private companies, the farmers made their first mistake, according to Dr. L. Jalpathi Rao, a senior agronomist in the Warangal Agriculture Research Centre, by abandoning the short-duration variety of cotton suitable for the low rainfall and shallow soil of Telengana. They planted RCH-2, a long-duration variety, suitable to areas with assured irrigation. The drought condition in the beginning and the erratic power supply compounded the problem of poor irrigation.

In 1994–1995 the total area under cotton cultivation in Warangal was 69,286 hectares, which increased to 100,646 hectares in 1996–1997. Commensurate with the increase in acreage was the increase in cotton arrival in the Warangal cotton market. In 1994–1995 the total arrival of cotton was 676,993 quintals, which increased to 13,38,330 quintals in 1996–1997. The increase in cotton production led to a decline in prices. In 1994–1995, the average price per quintal of cotton was Rs. 1,809, which went down to Rs. 1,618 in 1996–1997 (see table 9.3). However, there was no decline in the input cost per acre; instead the input cost in cotton has been increasing every year, says Dr. Jalpathi Rao.

In Warangal district the cotton crop basically replaced the crop rotation based on jawar (rabi) and green gram (kharif). Now these two crops have almost become obsolete. The acreage under the green gram–jawar sequence has shown a drastic decline in the last decade. In 1987–1988 the area under the green gram and jawar sequence was 143,500 hectares, which declined to 31,952 hectares in 1997–1998. Besides jawar and green gram, cotton has also replaced other oilseeds, especially sesame, groundnut, and castor. Today cotton is grown in 20–23 percent of the total cultivable area in Warangal. The total agricultural land of Warangal is around 4.5 lakh hectares, according to Dr. Jalpathi Rao.

In 1997–1998 the total area under kharif cotton was 99,150 hectares. Eighty percent of cotton farmers used RCH-2 (Research Cotton Hybrid-2); other varieties used by farmers were Somnath and Shaktinath of Nath Seeds, MECH-1, 12, and 13 of Mahyco Seeds, and Sunjiv of Indo-American Seeds. RCH-2 has been the most vulnerable to pest attack due to the compact planting or bushy planting of this variety. This variety grows horizontally and has a closed canopy, which protects pests because sunrays don't reach beneath the canopy.

In one acre, 450 grams of seeds (of any cotton variety) are sown. The cost is between Rs. 250 and Rs. 350 per 450-gram packet. However, when the farmer finds that not all seeds have germinated, he sows more, so about 500–600 grams of seeds are used in one acre. Since RCH-2 was very popular, farmers had to book this variety in advance; those who did not had to buy it on the black market at higher prices. However, the cotton failed due to severe pest attacks. The frequent sprays and spurious quality of pesticides used made them even more ineffective. Most farmers had to spend between Rs. 12,000 and Rs. 15,000 an acre on pesticides.

B. Ramanamma belongs to Gangapur village in Jadcherla in Mehboobnagar district of Andhra Pradesh. She and her husband cultivated twenty acres of leased land. Taken in by the marketing hype of seed companies, they replaced paddy with cotton. This proved beneficial at the beginning, but the crop demanded intensive irrigation, for which they took a loan of Rs. 50,000. The subsequent crops failed. Burdened with loans and accumulating interest, Ramanamma's husband committed suicide by consuming pesticide. Ramanamma and her son are today working as construction workers in order to survive.

The heavy investment in agrochemicals could not be recovered because the yield was much below the expected level; it did not even cover the input cost. Small farmers who had taken money and material on credit were driven into debt and then to suicide.

The agricultural season of 1998–1999 in the state of Andhra Pradesh echoed the experience of the preceding years. Facing incessant rains followed by drought, working hard for the whole year and not getting a reasonable price for their produce, unable to pay back loans from private moneylenders, farmers succumbed to suicide. Within Andhra Pradesh, more than 80 percent of total farmers' suicides occurred in the Telengana region of the state alone, with Warangal district sharing 40 percent of total deaths.

Farmers, lured by advertisements for cottonseeds, cultivated their lands with new varieties of cotton, including Navratan Ajith, Parry White Gold, and Bioseed. Mindful of the losses incurred during the past cotton crop, farmers have cultivated cotton with the utmost care. In spite of that, the adulterated seeds have destroyed thousands of acres of cotton crop in Parakala, Regonda, Atmakuru, Geisukonda, Sangyam, and Dharmasagar mandals of the district.

In Warangal district during 1998–1999, the extent of area cultivated by Navratan Ajith, Parry White Gold, Bioseed, and other varieties of cotton was around thirty thousand acres, which was spread across two hundred villages in twenty-seven mandals. It is believed that about six seed companies were successful in introducing these varieties in the villages through their field distributors.

Interestingly, the seed companies select their seed distributors from the village itself. These distributors are well-off farmers who operate large farms and influence decision making in their village. The films the seed companies show to farmers have been found to greatly influence their decision making about what type of seed to buy. Many farmers report that the boll size and the opened boll appeared very good in the films—however, they often could not get a single boll in their fields, and whatever bolls formed were shed by the plant without opening.

In the village Ulligedda Damera in the Atmakuru mandal of Warangal district, the whole village had planted a total extent of 150 acres with the Navratan Ajith variety of cotton in 1998–1999. Madarappu Ramesh, who had cultivated Navratan Ajith, invested a total of Rs. 10,000 to Rs. 11,000 per acre on his cotton crop. Nearly 70 percent of this investment was spent on chemicals and fertilizers. In the same village another farmer, Gudur Rajaiah, had cultivated three acres of land with the Navratan Ajith variety; he also incurred the same cost of cultivation for the cotton crop. His situation was worse than Ramesh's as he had a debt of Rs. 90,000 from the *arthies,* or private moneylenders, at an interest rate of 36 to 48 percent. These farmers and others came to know about this variety of seed from video films showed to them in their village. And almost all the farmers were in debt either to arthies shops or to landowners.

In another village, Pallarigudda in the Sangyam mandal of Warangal district, almost all the farmers had cultivated their fields with Parry White Gold (PWG). The standing crop was very robust but without any bolls on the plants. About 150 villagers had initiated complaints in the district consumer redressal forum at Warangal against the failure of PWG, demanding appropriate compensation from the company. Government officials visited the fields of farmers whose cotton crops had failed. The villagers also requested that government officials clear their spiraling debts.

Lack of Support for Agricultural Workers

In addition to the seed failure, in many mandals yellow-insect cotton pests had destroyed the entire standing crops in 1998–1999. Farmers reported that the agriculture department of the state had shown total negligence in disseminating the advice of scientists, which had resulted in the havoc caused by the pests. Moreover, farmers persuaded by the

Kottula Yakayya of Samudrala village in Station Ghanpur Mandal committed suicide in 1999. His family owns four acres of land. On two acres chilies were sown, and on two acres cotton was grown. Last year he borrowed a sum of Rs. 25,000 for cultivating cotton. With interest it totaled Rs. 60,000. Moneylenders started pestering him for payment of the interest. Not getting a proper price for his cotton in the market and unable to know how to clear the heavy debt, the farmer committed suicide by consuming insecticide.

Pacchikeyala Kameshwara Rao of Akinepalli village of Mangapeta Mandal, unable to bear his financial problems, committed suicide the same year. Insects intensely attacked his cotton crop, and the use of many insecticides could not stop their spread. His crop was completely destroyed. Receiving no advice from scientists and agricultural officers, he became despondent and consumed poison in the form of insecticides.

Lack of scientific advice from agricultural departments led to many more suicides, including those of Indala Ayilayya, Malotu Danja, Tallapalli Lakshamayya, and Pentala Odelu.

suggestions of pesticide sellers had bought at high cost inferior chemicals that could not reduce the pests' attack on the cotton crop. Also, the rate for cotton per quintal was not more than Rs. 1,500, which was not commensurate with the investment made on the crop.

Despairing over their investment losses and their inability to feed their families, farmers are committing suicide by consuming pesticide mixed in curd rice. The number of suicides reported during November and December 1998 was about fifteen. Of these suicides, most were of farmers forty and older. They left behind families who have no one to look after them. The story of two such farmers is given in the box above.

Discussions with seed and pesticide merchants at Warangal revealed that the seed companies provide a very high margin on their products and also they do not demand immediate cash payment from the seed and pesticide merchants. About 80 percent of these transactions are on a credit basis of forty-five to sixty days. The merchants pay the seed companies through postdated checks. In turn, the merchants sell the product on credit to the farmers, who are lured by the "helping hand" extended by these merchants. Since the farmers need not pay in cash, they can easily become trapped in increasing debt. Further, the same merchants who sell the seeds also sell the chemicals and fertilizers required for the crops. Getting everything on credit from one source makes the farmer vulnerable to every suggestion given by the merchant. In this way the farmer sinks into a marsh of indebtedness.

Seed and chemical companies operational in Warangal include Shaw Wallace, ICI, Rallis India, Monsanto, Saral India, Novratis, Nocil, and Bayer. The liberalization of the seed sector is an epidemic leading to suicides and high debt for the purchase of seeds, agrochemicals, and pesticides.

Growing Seed Scarcity

Globalization and privatization of the seed sector have eroded the seed supply of farmers and the public sector. Although the entry of private seed companies is justified on grounds of increasing farmers' options and choices, by making farmers look down on their own varieties as inferior and by eroding the capacity of the public sector, globalization has in effect created a seed famine.

There is a great mismatch in the number of seeds demanded by farmers and the seeds supplied. With this widening gap between demand and supply, the plight of farmers is getting worse. The demand for seeds of all crops has nearly doubled within six years. Public sector agricultural departments, state seed development departments, and oil-fed supply around 20 percent of the total requirement of seeds in Andhra Pradesh. Taking advantage of the deficit supply of seeds, private sector seed companies are providing spurious seeds.

Unavailability of seeds is also creating conditions of distress, causing farmers to resort to suicide. For instance, in the Rayalseema region of Andhra Pradesh, 50 lakh acres of groundnuts are cultivated. It is known to farmers that the groundnut crop in the kharif season can withstand ten to fifteen days without rain, a peculiarity utilized by the Anantpur and Kurnool regions, which cultivate this crop. Anantpur district requires 14 lakh quintals of groundnut seeds, but the government supplies only 1.12 lakh quintals. If farmers agitate for more, the government could provide a further 68,000 quintals, but even that would total only a meager 12 percent of the requirement. Thus, public-sector seed companies are unable to meet the demand, and the situation worsens every year. Taking advantage of these conditions and also the government's privatization spree, private seed companies are trying to reap benefits through selling unreliable seeds.

When seeds are unavailable, farmers either leave their lands fallow or are forced to change to other crops. Or worse: Gogoti Bali Reddy from Kuntalapalli village in Nallamada mandal in Ananthpur district succumbed to suicide due to having no seeds to sow.

In the agricultural season of 1999–2000, five lakh acres of rich fertile land were left fallow without any crop. The scenario is the same everywhere. In the ghat region (basically tribals), farmers were not able to raise their paddy nurseries due to lack of paddy seeds. Similarly, the subsidy available on seeds has been removed. During the cropping season of 2000–2001, the Department of Agriculture unearthed a racket operating in the distribution and sale of spurious banni cottonseeds. The farmers have so far planted fifty thousand acres of land with banni seed in the districts of Guntur and Prakasham.

Another aspect of seed scarcity is the precipitous rise of seed prices. Cottonseeds are now sold at double the price of the period of easy availability. This appears to be a deliberate strategy to create a market for the genetically engineered Bt cotton, which will be relatively higher priced.

Seeds, Pesticides, and Debt: The Intimate Nexus of Corporate Feudalism

In Warangal, land is easily available on lease because of the heavy migration of people from the villages to the city. Farmers with small landholdings often take land on lease to grow cotton, paying Rs. 1,800–3,000 per acre as an annual rent. Rajmalla Reddy of Atmakur mandal, for example, has forty acres of land, thirty-five of which he gives on lease every year for Rs. 1,800 per annum for each acre. Lands that have irrigation facilities fetch up to Rs. 3,000 per annum, said Mr. Reddy. Attracted by the prospect of getting rich overnight, peasants who lease land spend thousands of rupees on the pesticides and fertilizers necessary for conventional cotton cultivation. Besides contributing their own resources, middle and small farmers borrow money, paying high interest rates, from arthies, who also provide them with seeds, fertilizers, and pesticides on credit. These private moneylenders take on the role of "pest-management advisors," extend credits to farmers, sell spurious pesticides made by fly-by-night companies, charge higher than prevailing prices, and recommend the application of excessive doses of these pesticides.

The rise of moneylenders is a part of an emerging phenomenon of corporate feudalism. Withdrawal of low-interest credit has been a key element of the World Bank–led economic reforms. As cooperatives and rural banks close down, and public sector banks are privatized, rural credit dries up and farmers are pushed into borrowing from moneylenders. The failure of the private sector in Indian banking was what had ushered in the nationalization of banks in 1966. The prenationalization period had witnessed the growth of a banking system which, driven by profits, could not cater to the development needs of the nation. Credit was virtually inaccessible to the large masses of the rural and poor population. Lending policies were turned to the advantage of industrialists, with banks being under the control of industrial chairmen. Banking came to be controlled by a few communities, making it a family profession. The nationalization of banks was followed by a sharp increase in the number of bank branches. Consequently, employment shot up. Further, banking policies were tuned more to cater to the development needs of the nation, as priority-sector lending took precedence over profit-driven lending. In protecting the poor from the clutches of unscrupulous moneylenders, the nationalization of banks had succeeded in building up the productive base of regions and areas that would have otherwise remained neglected, through a number of projects and programs targeted particularly at women and other weaker sections of society.

The opening up of the banking sector to competition from private domestic and foreign banks has been accompanied by a reversal in these positive trends. For instance, there has been a fall in the proportion of credit received by the household sector, which had earlier received a relatively larger share of bank credit. Further, the incremental expansion during the postreform period for the household sector has not only been the smallest but is also smaller than expansion in favor of corporate enterprises. Similarly, the financial assistance sanctioned by the all-India financial institutions suggests that while disbursements of development financial institutions (DFIs) generally assisting large-scale industries expanded by 197 percent between 1990–1991 and 1994–1995, those of DFIs assisting small-scale and medium industries have risen by 62 percent only.[2]

Foreign banks are concentrated in metropolitan areas and completely absent in rural areas, while private banks are mostly concentrated in semi-urban areas. In the event of the nationalized banks giving way to private participants, it wouldn't be long before the rural areas are isolated from the financial scene. These trends are suggestive of a return to the prenationalization era, a demonstrated failure.

The private moneylenders are mostly pesticide dealers or shop owners. In Warangal there are thirteen thousand pesticide shops that distribute pesticides produced by ninety-three companies registered in Andhra Pradesh and also by about two hundred contraband units based in Maharashtra.[3] In each village there are five to eight shops. The shop owners and dealers get their supply of the stock from the pesticide companies on credit. So there exists a chain of credit system, and the shop owners are only the mediators. In reality, the farmers indirectly get the credit from the company itself. The interest rate varies from 36 to 60 percent per annum. Since the chemicals are easily available on credit, farmers have no hesitation in using it often, usually once a week, and at a high intensity. There is no government agency to finance the farmers, and bank loans are negligible. This has forced farmers to approach the private moneylenders.

The cotton farmers in Warangal spend about Rs. 1,500 on preparing their fields (especially on labor costs). The sowing period is June to July. In fields that are rain fed, sowing is delayed till it rains. One week to ten days after sowing the cottonseeds, farmers do the first spray of pesticides. This is done without ascertaining the existence of pests in their field. The first spray is considered crucial; it is believed that without it the crop will fail. However, the state government's Agriculture Department and the Agricultural Research Station, Warangal, have suggested to farmers the integrated pest management (IPM) strategy to control the pests through growing "trap crops" (such as a castor, marigold, and pheromone trap) in the field to see whether pests are actually present.

But farmers brainwashed into the miracle seeds/miracle spray culture do not pay heed to these suggestions, and within ten days of sowing, they start spraying their cotton fields with pesticides. Initially they use lower concentrations of chemicals. The chemicals that are used in the initial stage of spraying are Monocrotophos 36% EC, Dimethoate 30%, Oxydemeton-methyl, and so on. Mixing two chemicals is very common. In the first spray, only 250 milliliters is used in one acre of land. But from the second spray onward, 50 milliliters are added each time, and at one stage farmers end up using one liter of chemicals per acre. In one season, besides expenditure on fertilizers, labor, and seeds, cotton farmers spend Rs. 8,000 to 10,000 on pesticides alone. Pesticide is a major input in cotton. Once a week 300 to 500 milliliters of pesticide are sprayed per acre, and in one season (June–March) twenty-five to thirty-five sprays of pesticide are normal practice in Warangal. Among all the Indian states, the maximum use of pesticides is in Andhra Pradesh. A major portion of this is used in cotton and chili cultivation. Cotton is quite susceptible to a range of pests and diseases. In the 1980s, pesticide consumption in Warangal was less than Rs. 10 crores. But as the hybrid cotton cultivation picked up its momentum in 1985–1986, pesticide use also increased. In 1997–1998 the approximate sale of pesticides in Warangal district alone is Rs. 200 crores,[4] which is the highest in Andhra Pradesh, and almost 80 percent of this is used in cotton.

The pest problem is not new in the Telengana region; the farmers of this area have been facing this problem for the last three years. But in 1997–1998 the problem was very severe and the pests attacked almost all standing crops in the fields. There was a heavy loss of crops. However, the most affected crop was cotton. Cotton farmers were hardest hit because input cost in cotton was higher and the yield was not as expected. Earlier, cotton farmers used to get ten to twelve quintals of yield in one acre spread over four to five pickings. But in 1997–1998 they could hardly get four to five quintals. Some of the farmers could not get even that. The temptation of heavy returns on cotton had attracted small farmers, who had even leased land for growing cotton. Bandi Kalavathi, wife of Somaiah of Venkatapur village, had no land of her own but she had taken five acres of land on lease and in four of these she had planted only cotton. She had taken on Rs. 35,000 of debt from private parties. Bandi Kalavathi is one of the farmers who committed suicide due to the crop failure.

**Andhra Groundnut Crop Failure Drives Farmers to Suicide
Hyderabad, September 24.**

Close on the heels of suicides by Mehboobnagar cotton farmers
during April–May this year, death has once again begun to take a
heavy toll in the fields of Anantaput district, bordering Karnataka.
As many as seven farmers and two girls have committed suicide in
the districts during the last four days due to pest attack that almost
wiped out the entire groundnut crop in 3 lakh acres.

—*Hindustan Times* (New Delhi), September 25, 2000

In the cotton cropping season in 1997–1998, not a day passed after
mid-December 1997 without at least one farmer ending his or her life as a
consequence of the failure of the cotton, chili, red gram, and other crops
in Warangal, Karimnager, Medak, Rangareddi, and Mahabubnagar districts in the Telengana region and Kurnoor in the Rayalaseema region.

Incidentally, this was not the first time that such suicides had taken
place in Andhra Pradesh. In 1987 in the Guntur and Prakasham areas,
the cotton farmers faced a similar predicament, as did tobacco farmers in
other areas in subsequent years. Farmers were encouraged to shift from
their traditional self-sufficient cropping (of paddy and vegetables) to more
remunerative cash crops. But unlike the case with their traditional food
crops, total reliance on cash crops entailed a gamble, since fluctuations
in the market price affected their earnings. Moreover, their cultivation
involved huge expenditure on inputs like fertilizers and pesticides.

For the pesticides industry, the pests are a blessing in disguise, sustaining profit margins over the years regardless of the extent of crop
damage. The more the pest incidence, the more lethal the pesticide cocktail. Consequently, the insects became resistant to all kinds of pesticides.
Today controversial synthetic pyrethroids are also available. The pyrethroids are more expensive and are known to have a knockdown effect
on birds and other animals. They are also believed to be carcinogenic. No
sooner did the pesticides trade push in the pyrethroids than the insects
developed immunity against these fourth-generation pesticides. There are
twenty-eight known natural enemies of pests in the cotton fields. Nature

has provided enough protection for cotton through the abundance of benign insects, parasites, and predators available in the field: spiders, ladybird beetles, crysopa, wasps, rats, frogs, snakes, birds, and others. But the tragedy is that these parasites and predators are the first to be killed when pesticides are sprayed. With their natural enemies gone, pests are stronger than ever in the crop field. In Warangal the indiscriminate use of pesticides has reduced the bird population. When pesticides disturb nature's equilibrium, many of the little-known and previously insignificant pests of cotton, like white fly and *Spodoptera,* emerge as major pests.

There are more than fifty chemicals used in agriculture, and more than ninety companies are selling their products in Warangal district. There are several companies that are selling spurious and low-quality chemicals to which pests have developed resistance. As a result, farmers use higher concentrations and more expensive pesticides. Mixing two to three chemicals in order to combat pests has become a normal practice.

Besides pesticides, the cotton farmers also use fertilizers. In one season, about 150 kilograms of fertilizers, which cost about Rs. 1,500 to Rs. 2,000, are used in one acre. Every cotton farmer uses DAP and urea. Besides urea, they either use 17-17-17, 28-28-0, 14-35-14, 16-20-0-15, ammonia, DAP, and so on.

The two pests that attacked the cotton crop in 1997–1998 in Warangal were *Heliothis* and *Spodoptera*. Before these pests attacked cotton, the sucking and chewing pest white fly had attacked groundnut and chilies. In October–November, *Spodoptera* attacked the cotton crop. Though this is not a major pest for cotton, it heavily attacked cotton as well as groundnut, chilies, pulses, and others. The *Spodoptera* eats everything that is green—leaf, buds, flowers, and capsules. It is a voracious eater and moves in groups, attacking one field after another. *Heliothis,* on the other hand, eats only cotton capsules and buds. From morning to evening it remains under the soil and comes up in the evening for eating. That is the reason pesticide sprays don't affect it. In 1997–1998 farmers had to use poison baits to kill this pest.

Consequences of Overburdening Debts: Distress Sale of Kidneys

The ever-growing interest rates and the accumulating debts in Rentachintala mandal of Andhra Pradesh has led to the distress sale of kidneys by many farmers. The farmers are caught in a lose-lose situation; there is no

---

A. P. Farmers Sell Kidneys to Avoid Penury
Guntur, May 15.

Rentachintala, once again the hottest place in Andhra Pradesh, is back in the news, for an altogether different reason. At least 26 persons, all in their prime age, have sold their kidneys for pecuniary gains.

Pushed into the clutches of penury, the handful of small time farmers found an easy way out from their debt trap at the cost of their kidneys. The gravity of the situation can be gauged from the fact that at least 100 persons underwent fitness tests.

A "seller" Mr. Polli Reddy said he had no other option. "We owe thousands of rupees to the money lenders. They gave us loans to raise crops, mostly cotton and chillies. We could not reap a good crop in the 1st two years. The growing interest was draining our pockets."

—*Hindu* (New Delhi), May 16, 2000

---

way out of debt and no escape from humiliation at the hands of arthies and moneylenders/pawnbrokers.

The farmers here switched to cultivating chili, as usual driven by the lucrative returns. The investment during initial years was relatively low, as they were using native seeds, which are known for requiring less chemical intake. However, with the monoculture of chili cultivation spreading the damage, disease attacks increased, and every year the standing crops were affected. The loans the farmers subsequently took out after failed crops each year were used to sustain themselves, with whatever amount was left over put toward the next cropping season. Farmers went to pawnbrokers to get money to buy the necessary chemicals and sprays. The pawnbroker is a major actor who always comes out ahead in dealing with farmers. He "supports" farmers by providing loans at very exorbitant interest rates and also sells them chemicals from his shop. Individual loans may be small, but they accumulate over years and the farmers find themselves in heavy debt.

Once farmers are deep in debt there is no alternative available but to sell their land, agriculture implements, or homes. Recently, some farm-

> Bobba Venkat Reddy got deeper and deeper into debt as spurious seeds and chemicals ruined his crops year after year. Continually harassed by moneylenders, he heard of a broker who was helping farmers get money by selling their kidneys. This was a better option than suicide, and he took advantage of it. However, the surgery has left him weak and unable to work his farm. Because of the media coverage, moneylenders have refused to loan any more money to him and farmers like him.

ers in Rentachintala and surrounding areas like Gurazala have sold their kidneys in order to clear their outstanding debts with the pawnbrokers. Farmers from Rentachintala mandal who sold their kidneys are:

1. Durgyampudi Chinna Venkat Reddy
2. Dirsinals Narsi Reddy
3. Bobba Venkat Reddy
4. Siddhavarpu Poli Reddy
5. Peramlacchi Reddy
6. Kancharla Krishna
7. Narmala Krishna
8. Golle Ramaswami
9. Thai Narsaiah

Since the sale of farmers' kidneys came to light, the life of these farmers has become even worse. There is no support either from the government or from the village itself. These farmers are looked on as untouchables, and no one is coming forward to extend support to the deprived families.

The Karnataka Scenario of Farmers' Suicides

Crop Failures

In the 1999 cropping season, farmers at Harobanavalli village in Shimoga taluka reported that the 1001 paddy variety, which is very popular in the region, failed to perform in the second cropping season. Around ten farm-

In 1999, Gaddilingappa cultivated four acres of land with the C-71 variety of jowar supplied by Cargill. The company assured 20–25 quintals of yield. However, he got only 1 to 1.5 quintals. All the farmers who had taken up the seed variety went to the agricultural commissioner and senior officials in the agricultural department. An inquiry by the commissioner revealed that the seeds were meant for the kharif season, not the rabi. Farmers picketed Cargill at Bellary, after which Rs. 380 per acre was given as compensation. Farmers also demanded that twenty-eight tons of seed the company still had be destroyed.

ers reported that the paddy variety 1001 supplied by Rallis Company in this village failed.

The cost of the seeds is steadily increasing over the years. The problem with the 1001 variety was that despite regular applications of fertilizers and other chemicals, there was drastic reduction of yield. The farmers had been using the company seeds for a long time and therefore depended on the market for the seeds. Some of the farmers reported that the company cautioned the farmers for not using 1001 variety for a second time on their fields. They are apprehending this as a possible reason for the failure of the crop.

The case with the horticultural crop of chilies is also not good. This crop, though, has little to do with companies supplying seed, as seeds saved by farmers are mostly used. The farmers largely depend upon two major regions for supply of the seeds—one is in Karnataka itself called Baidagi, and the other is Guntur (Andhra Pradesh).

In the Bellary region, out of the thirty-five thousand acres of chilies planted during 1999–2000, around twenty-six thousand acres of the crop suffered total destruction. This amounts to nearly 70 percent of the area planted. Per-acre investment for the chili crops is between Rs. 16,000 and Rs. 20,000, of which the majority is on chemical sprays. The returns from the output were around Rs. 2,000 to Rs. 4,000 per acre. The failure of the crop was attributed to excessive rainfall and a subsequent attack of viral disease. This amounted to huge losses for the farmers, who have taken out loans from commercial banks at the rate of 30 percent per annum.

The number of suicides related to chili crop failure during 1999–2000 as reported by government agencies was around eight individuals; the figure reported by one concerned nongovernment group was nineteen individuals.

Another problem chili farmers face is storage facilities for their harvest. The government cold-storage facilities are becoming costlier, and farmers are not getting good prices despite holding the stock for longer periods. This is only adding to their costs, and the interests on their loans keep accumulating.

Seed Supply: Public versus Private Companies' Participation

All the agencies that are involved in seed production cater to the needs of farmers. These agencies need to provide quality seeds to these farmers by providing either certified seeds or labeled seeds. To sell certified seeds, agencies need to obtain a certificate from the state certification agency. They can also sell labeled seeds on their own. An analysis of all agencies providing various types of seeds shows that only public sector agencies pursue certification of seeds; those in the private sector operate without any proper certification.

In the kharif of 2000, an analysis of seed producers that have gone through seed certification in the state shows that more than 80 percent are public sector agencies, of which major players are Karnataka State Seeds Corporation and National Seeds Corporation of the State. Private sector participation in seed distribution in the state is around 12.28 percent of the total seed distributed. Interestingly, a detailed analysis of the sixty-five operational private seed companies that are registered with the state seed certification agency found that 88 percent of the companies are involved in the supply of cottonseeds, followed by maize, paddy, and bajra.

The Maharastra Scenario of Farmers' Suicides

Farmers in Yavatmal district in Vidhrabha have faced problems of cotton failure in the last few years despite favorable climatic conditions and uninterrupted supply of inputs. The yields have drastically decreased from a quintal to a few kilograms per acre over these years.

The crisis is very severe, and farmers are struggling for survival in the wake of the failure not only of cotton but also of other important crop seeds such as *toor* (pulses) and others. Till 1992, the majority of farmers

In Maharashtra, the people were growing millets, but agriculture departments working as extension workers for seed corporations advised them to stop growing millets and to start growing soya. Because they would get more money. They went in for soya. The farmers when growing millets were getting foliage that helped them to keep cattle, which produced dung to make the compost, which in turn, went back to the farmers' fields. Now when soya came to the market, the soya oil went to some factory, the soya cake was exported to USA for feeding pigs; the cattle had nothing to eat, the soil had no dung. It started to lose its fertility. The cycle started to work in the wrong way. The need of the hour is to look into the soil aspect of the farms. The concept of soil aeration is of utmost importance. Soil aeration is not taken into account by the western education. We forget about the air. The earthworm is one such type of organism that helps in soil aeration. GM crops endanger the soil component and the concept of living soil will vanish in the course of time if such crops are allowed. The need of the integrated approach to organic farming where the whole cycle of life is again rejuvenated.

—Dr. Sultan Ismail, leading earthworm ecologist

82 Maharashtra Farmers Committed Suicide This Year

As many as 82 farmers from Vidarbha and Marathwada regions of Maharashtra had committed suicide during the year after being overburdened by debts, The Revenue Minister, Mr. Narayan Rane told the Maharashtra Legislative Council today. The State Government had given financial assistance to 25 farmers.

—*Hindu* (New Delhi), July 21, 1998

were cultivating a basic normal hybrid (AHH 468) of cotton, which was fairly consistent and provided normal yield. The problem in this region started after 1992, when a new variety of cotton (CAHH 468) was introduced to the farmers in the region. The farmers noticed that the new hybrid, which has not been certified by the government, failed to perform well in spite of all the care taken by them. As reported by the farmers, the yield registered was almost negligible in subsequent years. These seeds were supplied to farmers by some seed companies they have trusted for years, including Nath, Aurangabad, Ajith, Jalna, and Sanjay. The government seed-selling outlets are supplying substandard seeds to the farmers. Some farmers have brought this to the notice of the authorities of these seed companies. For instance, *karadi* (Bhima) seeds (marketed by Mahabeej, Akola), which have been duly certified by the certifying agency, were found to be substandard.

The Punjab Scenario of Farmers' Suicides

Punjab—the highest contributor of grain to the national pool—enjoys the notorious distinction of having the highest rate of farmers' suicides among all the states. They started in 1990, but in 1997–1998 the problem became very severe, and today it has reached alarming proportions. Despite some government agency reports to the contrary, the alarming rise in rural suicides has become an acknowledged fact.

Initially the Punjab government was not ready to acknowledge that the suicides were occurring. But when the media reported the suicides, then the government formed committees to look into the matter. But unfortunately, these committees failed to pinpoint unremunerative agriculture, increased cost of production, and large-scale indebtedness as major reasons for the misery of farmers and instead concluded that the suicides are taking place due to family problems, intoxication, and other social reasons.

The study *Suicides in Rural Punjab* conducted by the Institute for Development and Communication, Chandigarh, in 1998, confirmed that there has been a distinct increase in the number of suicides in Punjab since 1998. In 1992–1993, suicides in Punjab increased by 51.97 percent; in 1993–1994 there was an increase of 14 percent; in 1994–1995 the increase was 57 percent. It notes with concern that suicide rates, that is, the number of suicides per lakh population, has been steadily increasing from 0.57 in 1988 to 2.06 in 1997 in Punjab.[5] It has been also observed that the percentage share of cultivator-farmers' suicides among total suicides in Punjab in 1991–1997 was 23 percent. In Sangrur district the percentage share of cultivator-farmers among total suicides was 50 percent. The suicide rate of cultivator-farmers in 1993 was 1.98, which increased to 4.49 by 1997. The

The state of Punjab covers an area of 50,33,000 hectares, constituting about 1.57 percent of the total geographical area of the country.

There are seventeen districts: Amritsar, Bhatinda, Faridkot, Ferozepur, Gurdaspur, Hoshiarpur, Jalandhar, Kapurthala, Ludhiana, Sangrur, Ropar, Mansa, Fatehgarh Sahib, Nawa Shahr, Moga and Muktsar, which have further been divided into 138 blocks comprising 12,795 villages.

The population of the state as per the 1991 census is 20.28 million. Out of this, 14,288,744 is the rural population, while the urban population is 59,93,225.

The state has been divided into three agroclimatic zones: the sub mountainous zone, central alluvial zone, and southern dry zone. The climate of the state is semihumid to semiarid.

Of the total geographical area of 50.33 lakh hectares, the net sown area in 1997–1998 was 42.04 lakh hectares, constituting approximately 84 percent of the total, as against the national average of 42 percent.

The gross cropped area is 78.33 lakh hectares, and the cropping intensity is 186 percent. The net irrigated area in Punjab was 40.21 lakh hectares in 1997–1998, and 96 percent of the gross cropped area is under irrigation. In 1997–1998 the average fertilizer consumption works out to be 167 kilograms per hectare in Punjab, compared to the national average of 73 kilograms per hectare.

The number of small and marginal farmers having up to 5 acres of land is about 499,510, 45 percent of the total landholding in Punjab. The average size of small and marginal farmers' landholdings is only .99 hectares.

study clearly admits that the rate of suicides of cultivator-farmers has been on the rise in Punjab since 1993. However, the chief minister of Punjab downplays this alarming situation. Replying to a question regarding these suicides during the June–July Assembly Session in 1998 (question number 1,087), he said, "During 1996, 1997 and 1998 there were only 8 suicides of farmers and agricultural labourers in Punjab. One in Tarn-Taran in Amritsar district in 1996 was a result of family dispute. The three in Sangrur were due to crop damage and indebtedness. Three in Bhatinda were

---

**Crop Failures Lead Punjab Farmers to Suicide
Chandigarh, April 20.**

About 80 cases of suicides by farmers and agricultural labourers reported from five villages of Sangrur district in the last four or five years could only be the "tip of the iceberg" as death stalks the rural areas of the Lehra and Andana blocks in the otherwise prosperous district of Punjab.

According to a former sarpanch, Mr. Jarnail Singh, and a jathedar, Mr. Mastan Singh, about 33 persons had been driven to suicide in Balaran village, while the figure was zero in the official records since 1994.

—*Hindu* (New Delhi), April 21, 1998

Fifteen years back, Paramjit Singh of Punjab cultivated chilies. Over the years, the cost of chemicals increased, and the yield declined considerably. Local moneylenders forced Paramjit Singh to sign blank papers in return for giving him loans and took over his land. When he could not repay his loans, they dragged him off the land. This was more than he could bear, and he committed suicide.

---

because of crop damage and indebtedness and one in Jalandhar due to crop damage by hailstorm."

Sangrur and Bhatinda districts reported the most suicides, with suicide rates of 12.08 percent and 6.24 percent respectively. It is also noticed that the share of noncultivators' suicides in these two districts is very high compared with other districts: 13.24 percent and 11.35 percent respectively. The districts of Mansa, Amritsar, Ferozpur, Gurdaspur, Faridkot, and Muktsar had moderate levels of suicide proneness in 1991–1997. All these districts comprise the cotton belt of Punjab.

Some analysts acknowledge the suicide phenomenon in Punjab but characterize it as a result of militancy. Countering this, Mr. Inderjeet Singh Jaijee, convenor of the Movement against State Repression, said, "If this were the case one would expect to find suicides limited to Punjab and that too to certain areas of Punjab such as the border districts. This is not the case. Lehra and Andana Blocks in Sangrur district have been identified as suicide prone area[s] and yet this part of Punjab was less affected by militancy. Likewise Haryana did not suffer the turmoil and disruption of militancy, yet debt related suicides are being reported from that state too."

The increased number of farmer suicides in Punjab can be understood in the context of growing distress in the agriculture of the state. The factors contributing to this state of affairs in agriculture are the decline in farmers' income from the farm, the increased cost of production, crop failures and crop losses, the monoculture of wheat-paddy cultivation, rising unemployment in the rural areas, and so on. According to an estimate of the Department of Economics and Sociology, Punjab Agriculture University, Ludhiana, "The annual surplus of small size farm is about Rs. 9,500 during 1993–84." It further estimates that "the best managed five hectares farm with standard field crop rotation, can earn barely an income equivalent to the average per capita income in Punjab." However, in 1999–2000 this could have declined further due to the increased cost of the production of principal crops in the state.

Green Revolution Is Not More Green

The prevalence of suicide among farmers in Punjab exposes the flaws in the much-vaunted Green Revolution. Today the village agricultural economy of Punjab is in crisis, and the living conditions of the farming community and farm laborers are in bad shape.

Mr. Prakash Singh Badal, the present chief minister of Punjab, has said, "Agriculture for most has become a pain in the neck. It is not profitable at all except for those who own ten acres or more. . . . What is in the hands of the state? Everything has been centralised. Prices of all inputs are controlled and fixed either by the industry or Union Government. The price of farmers' produce—wheat and paddy and most of the other produce—are fixed by the Centre."[6]

The farmers of Punjab are voracious users of inputs in their bid to enhance the productivity of agricultural crops. For example, Punjab consumes 10 percent of the fertilizers, 11 percent of the pesticides, and 55 percent of the herbicides used in the entire country. The same is true for other agricultural inputs like irrigation and use of farm machinery.

In Punjab the growth of agriculture is mainly confined to two crops, rice and wheat, and has reached its saturation point. The data of Punjab indicate that the productivity of rice was 4.89 percent from 1965–1966 to 1985–1986, and it declined to 0.58 percent from 1985–1986 to 1996–1997. The productivity of wheat has also declined from 2.79 to 2.14, and sugarcane declined from 3.40 to 0.28 in the corresponding period. The productivity of cotton increased to 1.63 from 1965–1966 to 1985–1986, but the total production of cotton in the state declined from 19.25 lakh bales in 1996–1997 to 9.41 lakh bales in 1997–1998 due to the pest attack and crop failures. However, in the 1980s, the Punjab Agriculture University, Ludhiana, made it abundantly clear that farmers with holdings of fewer than fourteen acres were fighting a losing battle for survival.

Essential factors such as soil health and water resources are being overstrained, and there are serious buildups of pests, diseases, and weeds. Pest has emerged as a very serious menace in Punjab due to monocultures. There is also no scope for further expansion of the area under cultivation or increase in the cropping intensity (which is at present at a very high level of 186 percent). The water resources of the state are being overexploited through the adoption of high-water-requiring cropping sequences and the use of high-yielding varieties.

Increased Cost of Production

The increase in the prices of inputs and labor has pushed the cost of production up during the last three decades (six times for wheat, seven times for cotton, and ten times for paddy). The increased cost of production has led to increased indebtedness among farmers in Punjab. Today 90 percent of farmers in Punjab are in the trap of debt.

To get an idea of the soaring cost of cultivation in Punjab, it is desirable to study the trends of the three major crops of the state: paddy, wheat, and cotton. The per-hectare cost of the cultivation of paddy in Punjab has increased by five times in a span of eighteen years, from Rs. 3,419.33 in 1978–1979 to Rs. 17,966.85 in 1996–1997. The cost of production per quintal of paddy has increased from Rs. 68.71 in 1978–1979 to Rs. 334.81 in 1996–1997 (see table 9.8).

However, despite the fivefold increase in the cost of the cultivation of paddy, there is no corresponding increase in its yield. The yield increased by just 2.17 quintals per hectare from 1978–1979 to 1996–1997. However, during this period there was a very significant increase in the use of fertilizer, insecticides, and machine labor in the paddy crop in Punjab and a very drastic decline in the use of animal labor. This also indicates that Punjab farmers have almost ceased doing any manual work on their farms, leaving it to either migrant laborers or machinery.

Fertilizer use in paddy increased from 163.85 kilograms in 1978–1979 to 195.49 kilograms per hectare in 1996–1997. During the same period, the total amount of insecticide, in terms of value, also increased, from Rs. 56.77 to Rs. 825.04, while the machine labor cost increased from Rs. 90.93 to Rs. 956.80. Unfortunately, the heavy use of machine labor had its impact on the animal labor on the farm, which declined from 21.89 pair hours in 1978–1979 to 1.99 pair hours in 1996–1997.[7]

The wheat crop has also shown a similar increasing trend in cost of production. There is a sixfold increase in the per-hectare cost of cultivation of wheat in Punjab, from Rs. 2,722.36 in 1978–1979 to Rs. 17,333.89 in 1997–1998. During the same period, the cost of production per quintal of wheat has also increased, from Rs. 108.57 to Rs. 411.97 (see table 9.9).

In wheat, the yield has increased from 22.61 quintals in 1978–1979 to 35.78 quintals per hectare in 1997–1998. In comparison to this, during the same period, fertilizer use also increased from 125.69 kilograms per hectare to 224.87 kilograms per hectare, the cost of insecticide per hectare increased from Rs. 0.95 to Rs. 428.83, and the cost of machine labor per hectare increased from Rs. 283.03 to Rs. 1,692.07. Due to the heavy use of machine labor, the animal labor declined from 45.44 pair hours in 1979–1980 to 2.47 pair hours in 1997–1998.[8]

Cotton is not untouched, either. There is a sharp increase in the cost of cultivation of cotton. In 1975–1976 the cost of cultivation was Rs. 2,154 per hectare, which increased to Rs. 19,497 per hectare in 1996–1997, a more than eightfold increase. Obviously, the cost of production per quintal has also increased in this period, from Rs. 225.95 to Rs. 1,703.04 (see table 9.10).

In cotton also there is no significant increase in yield in Punjab despite heavy use of fertilizers and pesticides. In 1975–1976 the yield of cotton was 9.11 quintals per hectare, which increased to only 10.93 quintals in 1996–1997. In comparison to the yield, the cost of insecticide use increased from

Rs. 51.99 to Rs. 2,401.49, and the fertilizer cost increased from Rs. 189.83 to Rs. 776.11 per hectare during the same period.[9] In 1999–2000 the total consumption of pesticides in Bhatinda alone was about 941,671 liters. Out of this, 90 percent was used only on cotton.

Cotton is a major crop in the southwestern districts of Punjab such as Bhatinda, Faridkot, Mansa, Moga, Muktsar, and Sangrur, which accounted for 13 to 20 percent of the national cotton production. But in the last few years there was a sharp decline in cotton production. The major constraints in the cotton crop include the inadequate availability of certified cottonseed, waterlogging in some pockets of the cotton area, and bad weather conditions during the cotton season. In Punjab about 80–85 percent of the area under this crop is covered by American cotton (*Hirsutum*) and the remaining area is under desi (*Arboraum*) (see table 9.12). Among the prominent varieties of American cotton are LHH-144, Fateh, F-1378, LH-1556, F-1054, F-846, and LH-900, and the desi cotton varieties are LDH-II, LD-491, LD-327, and LD-230.

In the last few years there has been a drastic increase in the population of sucking pests and aphids, jasids and bollworm, particularly American bollworm. The farmers are following a dangerous trend of spraying a cocktail of pesticides in hopes that one or another chemical in the mixture will control the pest.

One significant change occurring in the last couple of years in the cotton cultivation in Punjab is the increase in the area of hybrid cotton. It has increased from 10,200 hectares in 1998–1999 to 76,800 hectares in 2000–2001. This is a disturbing trend that will further escalate cultivation costs for farmers and will promote a very intensive use of pesticides as we have witnessed in Andhra Pradesh.

However, hybrid seeds by their very nature are high-risk seeds under high-input conditions affordable only by rich farmers. They may give good yields, but for resource-poor farmers, they translate into high risks and high debts. Also, hybrid seeds are highly pest prone and therefore need frequent pesticide applications. Pesticides create new pest problems as well as environmental and health hazards. Pesticides fail to control pest whether or not they are spurious. With the increase in the area of hybrid cotton, pest attacks will further increase and will create more problems for Punjab farmers.

Due to the pesticide treadmill, farmers borrow money on credit to buy pesticides. In 1999–2000 also, as the bollworm attacked cotton plants, farmers started taking out more loans to buy pesticides and insecticides to save their crops. Although Agriculture Department authorities maintain that there was only a mild attack of bollworm in the cotton belt, cotton growers of about twenty villages in the Talwandi Sabo block of Bhatinda pointed out that the attack was alarming. Sikander Singh of Bhai Bakhtaur village says, "*Sundi* (bollworm) has badly hit the crop. If the bollworm were not killed at this stage, it would kill the cotton growers." He was not able to repay last year's loan for buying pesticides and insecticides for spraying on the cotton crop, and this year he had to take out another loan for the same purpose.[10] Another farmer, Mr. Resham Singh, said that for the past six years he had been growing cotton and suffering losses. Every year he had been taking out loans to meet his agricultural and social needs, and now he was neck deep in debt.

Cotton farmers point out that bollworm has become resistant to insecticide and pesticides. If the government fails to take any action and the farmers fail to adopt alternative methods of pest control and integrated management of pests, Punjab might again witness a spurt in the numbers of farmers' suicides.

For the economic survival of small and marginal peasants, it is imperative to shift away from the pesticide treadmill that is pushing farmers into debt and suicide. The farmers are being forced into intensive industrial agriculture that is leading to the loss of their money, their land, and their lives.

Extensive Crop Failure

Besides the pest attack, another menace faced by farmers in Punjab is extensive crop failure and seed failure. Many analysts have attributed farmers' suicides in Punjab to crop failure and seed failure. Professor Gopal Iyer has acknowledged this fact in his report on suicides in Punjab. He says that "Punjab has also experienced substantial crop loss in cotton consistently during 90's and there was a major crop loss during 1998 Kharif. This fact has been adequately acknowledged by the Punjab Government in its report submitted to the Central Government for compensation to Punjab farmers due to crop loss for Kharif in 1998. The untimely rain in the third week of September and again from October

15 to 18, 1998 caused extensive damage to standing and harvested crops in Punjab."[11]

Farmers have also suffered huge losses because of seed failures. In the 1999–2000 cotton season, too, several instances of crop failure have been noticed. In the Jagaram Tirath village of Talwandi Sabo block, district Bhatinda, the Kohinoor variety of hybrid cotton is performing very poorly. Most farmers who have sown this variety are not happy with this, and they are now mentally prepared to face a total failure of this crop. Some realized this in the beginning and replanted the same field. Mr. Gurcharan Singh (son of Mehr Singh) and Mr. Gurdeep Singh Sarpanch had to plant again when their Kohinoor seeds did not germinate well.

Similarly, Mr. Mahinder Singh, son of Mangal Singh of Jagram Tirth village, also sowed Kohinoor hybrid cotton in seven acres. It is an early variety, as claimed by the company, but very few plants had given flower before mid-August, after 120 days of planting. The farmers said that by this time the bolls should have been ready. When the villagers went to the dealer to complain, he said that the bolls would come and that the same variety was doing well in Rajasthan, knowing full well that no farmers would go there to investigate. According to farmers, in more than twenty villages of Moud Mandi, Talwandi Sabo, Rama Mandi, and Mansa Mandi about 50 percent of farmers had planted Kohinoor cotton, but in all these villages it showed signs of failure, as reported by farmers of Jagram Tirth village.

The loss of their crop is a curse for indebted farmers, and in extreme cases they commit suicide.

Cropping Pattern: Trends toward Monoculturism

In Punjab, the cropping pattern shows a trend toward monoculture. Farmers are abandoning the cultivation of diverse crops, for example, pulses, bajra, jowar, and oilseeds and getting trapped into the paddy-wheat combination. This is one of the major reasons for farmers' declining productivity and income. They are now dependent on the market for their day-to-day requirement of pulses, oilseeds, and vegetables. Though Punjab is known for being the "food basket of the country and the granary of India," it is not bringing prosperity to its own farmers. The paddy-wheat combination in Punjab is wiping out agricultural diversity.

The area under rice has increased from 227,000 hectares in 1960–1961 to 2,519,000 hectares in 1998–1999, an elevenfold increase. The area under wheat increased from 1,400,000 hectares in 1960–1961 to 3,338,000 hectares in 1998–1999, while the area under cotton increased from 446,000 hectares in 1960–1961 to 724,000 hectares in 1997–1998 but declined to 475,000 hectares in 1999–2000 due to crops failures in the last few years. But in 1999–2000 the area under cotton again increased, to 550,000 hectares in Punjab.[12]

However, the area under pulses in Punjab has decreased drastically, from 903,000 hectares in 1960–1961 to 78,000 hectares in 1998–1999, more than a tenfold decrease. In the same period gram went down from 838,000 hectares in 1960–1961 to 132,000 hectares in 1998–1999, which is more than a sixtyfold decline. The area under maize went down from 327,000 hectares to 154,000 hectares in the same period. Area under oilseeds has also decreased, from 185,000 hectares to 158,000 hectares. Area under millets and coarse grains has also declined. In the case of bajra and jowar, the decline is very sharp, from 123,000 hectares to only 4,000 hectares and 17,000 hectares to nil, respectively, during the period from 1960–1961 to 1998–1999.

It is true that Punjab, comprising only 1.57 percent of the geographical area of the country, produced 19.3 percent of wheat, 9.6 percent of rice, and 8.4 percent of cotton of India's total produce during the year 1997–1998, and it contributes 40–50 percent of rice and 50–70 percent of wheat to the central pool. However, the increase in the area of wheat and rice has shifted the whole cropping pattern of Punjab from diversity to monoculture, and quite obviously the shift to monoculture would register an increase of monoculture output but a drastic decline in the output of the diverse crops.

The production of pulses has decreased from 709,000 tons in 1960–1961 to 50,000 tons in 1998–1999. Similarly, the production of oilseeds, millets, and maize has also decreased in Punjab due to the spread of monocultures of wheat and rice. This shift has left farmers with no option except to hope that they would get better yield next year. With that hope they are getting trapped on the treadmill of fertilizers and pesticides and keep sinking further into the swamp of debt and humiliation.

Their profit from agriculture has declined while their household expenditures have been increasing. Gone are the glory days of the 1980s and early 1990s, when their income had increased substantially due to the introduction of HYVs, good returns from cotton, and government-supported subsidized inputs to encourage the Green Revolution. Today the Green Revolution is no longer green. Neither are the HYVs performing a miracle of instant increase in yield; cotton has been failing in the last few years, and the government has been withdrawing the crutches of subsidy. Now to feed their farms with chemical fertilizers and pesticides and to feed themselves and their families, farmers are succumbing to the control of private moneylenders, tractor agencies, and seed, fertilizer, and pesticide dealers. Their burden of debt is increasing every year.

Reckless Mechanization of Agriculture

Though agriculture in Punjab is undergoing a severe crisis, there is no sign of a decline in the sale of farm machinery. The farmers of the state have been suffering due to the high cost of input-intensive agriculture. Table 9.15 on the increase in agricultural implements in Punjab from 1995 to 1999 validates this. In Mansa district alone, which is a very backward district of Punjab and a suicide-prone area, too, the total number of tractors of all brands sold every year is around twelve hundred, according to Mr. Kishor Chand, manager of Amar Tractor Agency.

Agricultural experts of Punjab blame tractors for the indebtedness of farmers. The tractor has become a status symbol for many farmers. At present, there are about four lakh tractors in Punjab. Even farmers with five to six acres of land buy tractors in this state. This has given rise to a secondhand market of tractors, and once a week, the tractor *mela* (market) is held in more than fifteen different places in Punjab. But this only highlights the bad state of affairs in Punjab, as farmers are selling their tractors to pay their debt or meet social obligations. Farmers buy a new tractor on loan and within a month resell it in the market at Rs. 50,000–60,000 less than the price they paid. This phenomenon is very prevalent among distressed farmers in Punjab. The reason for such resale, as acknowledged by some of the farmers, is to repay loans borrowed from local arthies.

Increased Farmers' Suicides

Suicides in Punjab have reached alarming proportions in recent years, especially in the southern districts, also the main cotton zone of this

Crop Failure and Mounting Debts Drive Punjab Farmer to Suicide
Bhatinda, Oct. 2.

Reeling under heavy debts and disappointed over the decay of his crop, Mohinder Singh (30), a farmer of Nat Bagher village, about 35 km from here, allegedly ended his life consuming pesticide. He has left behind a 27-year-old wife and three children.

Mohinder's uncle told the *Indian Express* that he (Mohinder) owed Rs. 2 lakh to a commission agent and money lenders. He has taken nine acres of land on contract at the rate of Rs. 7,000 per acre.

His cotton crop on seven acres was destroyed by American bollworm and other pests. Another farmer, Jarnail Singh said that about 90% farmers of the village were under debt owing to the bad crops for the past five years.

—*Indian Express* (New Delhi), October 3, 1998

Two More Farmers Commit Suicide in Punjab

In yet another case of debt and crop failure deaths, two farmers of Bir Khurd village in this district allegedly committed suicide by consuming pesticides. Bikkar Singh (39) and Baldev Singh (42) ended their lives on October 19 and October 16, respectively. Both of them were deep in debt.

—*Indian Express* (New Delhi), October 25, 1998

state. As reported by Dr. Gopal Iyer and Dr. Mehar Singh Manick of the Department of Sociology, Punjab University, the reason for farmers' suicides in Punjab is mainly high indebtedness. "Indebtedness among the farmers and farm labourers in Punjab has reached epidemic proportions. Landless agricultural labourers, small and marginal farmers are more vulnerable than large farmers. Large farmers are able to sell portions of their holdings to pay off debts, which acts as a buffer. The major thrust of the small and semi-medium farmers is to borrow primarily for agriculture and marriage purposes. The lending agencies not only pressurise the farmers to clear the outstanding loans but also humiliate them. They experience loss of prestige and are forced to commit suicide."[13] High indebtedness is followed by constant pressure from lending agencies to repay the loan, an important factor influencing farmers to commit suicide. Another important factor is family members' resistance to selling land to clear the debts, culminating in the suicide of one or more family members. In the Chek Ali Sher village in Mansa, three members of one family—a father and two sons—committed suicide when a moneylender claimed his title over their land.

Social Reasons

In fact, this culture of committing suicide to escape indebtedness and the social stigma of being financially broke started in Punjab a few years ago. Small and marginal farmers are opting for commercial crops such as hybrid cotton on a large scale and making huge investments in anticipation of a good return. To meet heavy investment demands, farmers take out private loans at a very high interest rate, 2 percent to 3.5 percent per month. This has given rise to several other social problems among cultivators' families in Punjab. According to psychiatrists in Punjab, the debt trap has led to an increase in the consumption of intoxicants as well as matrimonial and family disputes. Most farmers are very "status conscious."

Most Punjab farmers have insufficient income to maintain themselves as the expenses of the community rise. Many farmers who have lost their land, ashamed to work as laborers in their own village, migrate to cities in search of such jobs. In the cities they compete with migrant laborers, who are preferred by landlords because they are cheaper and better behaved than local laborers. Moreover, subsidiary occupations of the farming community such as animal husbandry, poultry keeping, beekeeping, and fisheries are also running at a loss. In most cases these businesses were started on loans that the entrepreneurs now find difficult to repay. In addition, landholdings are fragmenting into smaller and smaller parcels because of rising population and the disintegration of the family.

Education has been totally neglected in Punjab villages. There are schools in which as many as three hundred children share a single teacher. The literacy rate is lowest in the Mansa–Sangrur districts, where suicide has been a large-scale problem in the last few years. The arthies take advantage of the farmers' illiteracy; even after a farmer's debt has been totally discharged, they normally do not delete the farmer's name from their registers. There is a saying in Punjab, according to Subah Singh of Jagaram Tirth village, Talwandi Sabo, Bhatinda: "If a farmer takes a loan from a commission agent, it will never be over till his death." The situation is further exacerbated by the floods every year in this region.

Due to increased rural indebtedness in certain villages, all lands are encumbered. The farmers want to sell the land, but there are no buyers. The land price has come down drastically.

Credit Facilities to Farmers

Farmers in Punjab are borrowing from various credit sources/agencies. The main agencies that are financing the credit needs of farmers in the state are cooperative credit institutions like primary agricultural credit societies and primary land development banks, commercial banks, and regional rural banks and also the informal sector credit agencies like commission agents (arthies) and moneylenders.

Borrowing for financing the current farm expenses is on a short-term basis, normally for a crop season, and these loans are repaid (fully or partly) through the sale proceeds at the end of the season. Fresh loans are taken out to finance the working capital requirement of the next cropping season: a never-ending vicious cycle. The credit advanced to the farmers of Punjab increased six times between 1990–1991 and 1998–1999 (see table 9.16).

A formal credit agency lends money to farmers by registering their land as security in its name. Similarly, banks provide loans against the security of land. Once the loan is forwarded to the farmer, these agencies ensure that the farmer does not apply for a loan from any other bank by putting their stamp on the papers.

Over the years, banks and other financial lending agencies have changed their methods of recovering their losses from loans to farmers. Once the stipulated date of repayment is passed, representatives from the banks go through the village announcing the auction of the land on a loudspeaker. This method of auction, according to farmers, is being done to humiliate them as well as to terrify other farmers so that they make their payments on time. The three acres of land of Mr. Roshan Singh of Bhai Bhakhtuar village of Maud block of Bhatinda was auctioned in this way by the bank.

All farmers, irrespective of their own investment through loans, take out the meager loan of Rs. 2,000 provided by the bank. According to a study conducted by Dr. Shergill of Punjab University, the total debt on the farmers of the Punjab state is about Rs. 5,700 crore.[14] This debt is about 70 percent of the net domestic product originating in the state in a year. In other words, three-fourths of one year's total agricultural income of the state has to be paid if the total amount of debt is to be liquidated. However, to freeze the annual recurring interest charged on the total debt, about 13.2 percent of the total farmland area of the state will have to be mortgaged by the farmers. Seventy percent of the farmers are unable to repay their loans. The Punjab scenario is distressing—farmers must sell their land, tractors, and cattle at throwaway prices to meet their debt commitment. The cash expenditure of the farmers has been steadily growing, which has resulted in a continual decline in the net surplus generated from the production of their crops.

Loans through government agencies in the 1980s and early 1990s used to be waived by the government. But now it is a different scenario because the loans are being taken mostly from private moneylenders. According to Mr. Rudlu Singh, a farmer member of the BKU Ekta, Mansa, there are about twenty-four thousand commission agents in Punjab who charge compound interest for loan money, which is doubled in a short period of three years, three months, and nineteen days.

The arthies copy the formal credit institutions and register the land of the borrowers in their name as a security. When a farmer borrows a big amount from the arthies, he registers his land for the same value. If the borrower fails to repay the loan, he loses his land. But sometimes arthies give the land to the owner to cultivate as a tenant and not as owner. Due to social stigma and shame, the victim farmer never tells others in the village that he has become landless. About Rs. 8,000 crores of arthies' money is floating in the market in Punjab. These arthies pay no income tax on this amount. There is a total of 12,560 villages in the state of Punjab, and on average two arthies operate in each village and control the village's finance and economy, according to the BKU Ekta.

If the arthies fail to get their money back from farmers, they take away tractors, trolleys, and grains and sometimes occupy the house and lands of the defaulters. Mr. Mange Ram of Mansa Mandi took away the tractor and Rs. 82,000 from a farmer, Mr. Mahinder Singh, son of Mr. Arjun Singh, of Burj Tilam village in Mansa district for not repaying a debt of Rs. 3 lakhs to him. To pay the money the farmer had to sell his land. A farmer in the village Jattan Khurd in Mansa district had taken a loan of Rs. 65,000 and could not repay due to successive crop failures. The commission agent took away thirty-five to forty quintals wheat, the annual ration, lying in his house as well as his tractor with trolley. Even with the intervention of the BKU Ekta, the farmer got back only his wheat.

There are several farmers in Bhai Bhakhtawar village in Maud Tehsilin Bhatinda district whose land has been seized by arthies. Among others are Jagseer Singh (son of Jaggar Singh), Bant Ram Vpeywala, and Nichatar Singh. According to Dr. H. S. Shergill, "In 1997 farmers borrowed a whopping Rs. 3119 crore. Sixty one percent came from traditional commission agents. Here interest rates are between 24 to 30 percent. Cooperative could manage just 34 percent; the rest—a meager 4 percent—came from commercial banks."[15] This situation is particularly detrimental to small farmers as interest rates are determined by the size of the holding—the smaller the holding, the higher the interest. Such exploitation by the commission agents and the burden of debt are forcing several farmers in Punjab to commit suicide, and nobody in the government seems to be paying any attention. Unfortunately, these suicides are rarely reported to the police. The discrepancy between the actual figure (collected by activists and farmer unions) and the official figure is explained by the fact that many suicides go unreported; official figures are invariably lower than the real ones. There seem to be unanimous agreement among villagers in rural Punjab not to report these deaths to the police as suicides. The villagers justify this by stating it will avoid "desecration of the dead body during postmortem examination and associated harassment by the police."

The most common method adopted by farmers for suicide is drinking pesticides and agricultural fumigants, which are available in abundance. Hanging, drowning, self–immolation, and throwing oneself in front of an oncoming train are also resorted to by some farmers.

About 150 cases of suicide by farmers and agricultural laborers have been reported in the last four to five years from the Lehra and Andana blocks of Sangrur district. In a single village, Dhindsa of the Lehra block, in the last five years more than fifteen farmers have committed suicide due to crop failure and increased debt.

In 1999–2000 suicides by farmers continued in Punjab due to acute indebtedness, exploitation of commission agents, and crop failures. Mr. Tirth Anok Singh of Jagaranm village was in debt for Rs. 1 lakh, which he borrowed from an arthi. He also bought a tractor (Mahindra 256 DI) against his lands and was in debt for Rs. 2.5 lakhs to the State Bank of Patiala. But one month after the purchase of the tractor the arthi took it away. Mr. Singh left his house the day his tractor was seized and never returned. His son Mr. Pretem Singh said that his father might have committed suicide. The arthi sold the tractor for more than Rs. 2 lakhs but kept all the money; nothing was given back to Mr. Singh's family after deducting the loan money. Neither was any paper given to the victim's family after the sale of the tractor. The family continued paying interest to the bank; otherwise they would lose the land, because farmers' lands are registered in the name of the loaner bank. In April 2000, Mr. Sadhu Singh, aged forty years, of Dhindsa village, Mwonak Tehsil of Sangrur, committed suicide. In 1998 he took out a loan of around Rs. 35,000 from a commission agent. In two successive years his crop failed. He had also taken land on lease against his wife's jewelry. After his death, the owner of the land kept the jewelry and gave the land to the commission agent. In August 2000 two landless laborers, Mr. Surju, son of Chand, and Mr. Sukhdev, son of Preetam, of Dudian village under Mwonak Tehsil of Sangrur district committed suicide because they were not able to repay their debts.

Conclusion

India has once before been colonized through cotton. From being the biggest producer of cotton and cotton textiles, India was converted into the biggest market for textile produced by the British industry.

Today cotton colonization is not restricted to cotton textiles but goes deeper, into the colonization of cottonseeds. From being the country of origin and the center of diversity, India is being rapidly reduced to dependence on imported cottonseeds.

Freedom from the first cotton colonization was based on liberation through the spinning wheel. Gandhi's use of the charkha and the promotion of khadi were both forms of resistance to the British monopoly on cloth and reminders that it was in our hands to make our own cloth again.

Freedom from the second cotton colonization needs to be based on liberation through the seed. Indigenous seeds are still available in large parts of India. Organic cotton is promising to become a major route to prosperity for farmers in marginal and rain-fed areas. The freedom of the seeds and the freedom of organic farming are simultaneously a resistance against the monopolies of corporations like Monsanto and a regeneration of agriculture that will bring fertility to the soils and prosperity to the farmers. The seeds of suicide need to be replaced by seeds of prosperity. And those seeds should be in the hands of our farmers, not in the control of corporations.

Notes

1. *Observer,* June 8, 1999.

2. Sudhir Shetty, *Alternative Economic Survey* (1996).

3. Chakrabarti Asish, "Pesticides, Moneylenders Play Havoc with Andhra Farmers," *Farm Digest,* February 1998.

4. Ibid.

5. *Suicides in Rural Punjab* (Chandigarh: Institute for Development and Communication, 1998).

6. *Tribune,* May 15, 1998.

7. *Cost of Cultivation of Principal Crops in India, 1991, 1996 and February 2000* (New Delhi: Directorate of Economics and Statistics, Ministry of Agriculture, 1991).

8. Ibid.

9. Ibid.

10. *Tribune,* August 17, 2000.

11. Gopal Iyer and Mehar Singh Manick, *Indebtedness, Impoverishment and Suicides in Rural Punjab* (Delhi: Indian Publishers, 2000).

12. *Agricultural Statistics of Punjab on the Eve of New Millennium* (Punjab: Statistician Department of Agriculture, 2000).

13. Iyer and Manick, *Indebtedness, Impoverishment and Suicides in Rural Punjab.*

14. H. S. Shergill, *Rural Suicides and Indebtedness in Punjab* (Chandigarh: Institute for Development and Communication, 1998).

15. Ibid.

10

Seed Freedom— What Is at Stake

Seed is not just the source of life. It is the very foundation of our being. For millions of years, seed has evolved freely to give us the diversity and richness of life on the planet. For thousands of years farmers, especially women, have evolved and bred seed freely in partnership with each other and with nature to further increase the diversity of that which nature gave us and adapt it to the needs of different cultures. Biodiversity and cultural diversity have mutually shaped one another.

Today, the freedom of nature and culture to evolve is under violent and direct threat. The threat to seed freedom impacts the very fabric of human life and the life of the planet.

Seed keepers, farmers, and citizens around the world have joined together as the Global Citizens Alliance for Seed Freedom to respond to this seed emergency and to strengthen the movement for the freedom of humanity. The Global Citizens Alliance for Seed Freedom is the start of a global campaign to alert citizens and governments around the world of how precarious our seed supply has become and, as a consequence, how precarious our food security has become.

Seeds are the first link in the food chain and the repository of life's future evolution. As such, it is our inherent duty and responsibility to protect them and to pass them on to future generations. The growing of seed and the free exchange of seed among farmers have been the bases of maintaining biodiversity and our food security.

Navdanya was started twenty-five years ago to protect our seed diversity and farmers' rights to save, breed, and exchange seed freely, in the context of the emerging threats of the TRIPS (Trade Related Intellectual Property Rights) Agreement of the World Trade Organization, which opened the door to the introduction of GMOs, patents on seed, and the collection of royalties. A Monsanto representative later stated, "In drafting these agreements we were the patient, diagnostician, physician all in one." Corporations defined a problem—and for them the problem was farmers saving seed. So they offered a solution, and the solution was the introduction of patents and intellectual property rights on seed, making it illegal for farmers to save their seed. Seed as a common good became a commodity of private seed companies, traded on the open market.

Today, the threat is even greater. Consider the following:

- The last twenty years have seen a very rapid erosion of seed diversity and seed sovereignty, and the rapid concentration of control over seed by a very small number of giant corporations.
- Acreage under GM corn, soy, canola, and cotton has increased dramatically.
- Besides displacing and destroying diversity, patented GMO seeds are also undermining seed sovereignty, the rights of farmers to grow their own seeds and to save and exchange seed.
- In countries across the world, including in India, new seed laws are being introduced that enforce compulsory registration of seed, thus making it impossible for small farmers to grow their own diversity, and forcing them into dependency on giant seed corporations.
- Genetic contamination is spreading—India has lost its cotton-seeds because of contamination from Bt cotton, and Mexico, the historical cradle of corn, has lost 80 percent of its corn varieties, and these are but two instances of the loss of local and national seed heritage.
- After contamination, Biotech Seed Corporations sue farmers with patent infringement cases. More than eighty groups came together recently in the United States and filed a case to prevent Monsanto from suing farmers whose seed had been contaminated.
- As farmers' seed supply is eroded and farmers become dependent on patented GMO seed, the result is indebtedness. Debt created by Bt cotton in India has pushed farmers to suicide.

- India has signed a U.S.-India Knowledge Initiative in Agriculture, with a representative of Monsanto on the board, and states are being pressured to sign agreements with Monsanto. An example is the Monsanto Rajasthan memorandum of understanding (MOU) under which Monsanto would obtain intellectual property rights on all genetic resources as well as research on seed carried out under the MOU. In a campaign led by Navdanya and a "Monsanto Quit India" Beeja Yatra (Seed Pilgrimage), relentless protests by farmers forced the government of Rajasthan to cancel the MOU. Monsanto influence on the U.S. government and the joint pressure of both on governments across the world are major threats to the future of seed and the future of food.
- Wikileaks exposed the U.S. government's intentions to proliferate the use of GMOs in Africa and Pakistan. Pressure to use GMOs imposed by U.S. government representatives is a direct effort to support giant biotech business and to expand their markets.
- For the ballot initiative on GMO labeling in the United States, corporations led by Monsanto are spending millions of dollars to prevent citizens from exercising their right to know and right to choose.

These trends demonstrate a total control over the seed supply and a destruction of the very foundation of agriculture. The disappearance of our biodiversity and of our seed sovereignty is creating a major crisis for agriculture and food security around the world. We are witnessing a *seed emergency* at a global level. Determined action is called for before it is too late.

The Assault on Seed

A reductionist, mechanistic science and a legal framework for privatizing seed and knowledge of the seed reinforce each other to destroy diversity, deny farmers innovation and breeding, enclose the biological and intellectual commons, and create seed monopolies.

Farmers' varieties have been called landraces, primitive cultivars. They have been reduced to a "genetic mine" to be stolen, extracted, and patented. Not only is the negation of farmers' breeding unfair and unjust to farmers, but it is also unfair and unjust to society as a whole.

Industrial breeding has been based on strategies to sell more chemicals, produce more commodities, and make more profits. The high-yielding varieties of the Green Revolution were in reality high-response varieties, bred to respond to chemicals. Hybrids are designed to force the farmer to the market every season, since they do not breed true. "Yield" focusing on the weight of a single commodity is an inappropriate measure. Commodities do not feed people—they go to producing biofuel and animal feed. Quantity empty of quality and weight empty of nutrition do not provide nourishment. Beginning with the false assumption that farmers' varieties are "empty," industrial corporate breeding gives us seeds and crops that are not only nutritionally empty but loaded with toxins.

The rendering invisible of the diversity that seed farmers have bred began with the so-called Green Revolution. The Green Revolution narrowed the genetic base of agriculture, encouraging monocultures of rice, wheat, and corn. Varieties bred for response to chemicals were declared miracle seeds and high-yielding varieties.

Industrial breeding has used different technological tools to consolidate control over the seed—from so called HYVs to hybrids to genetically engineered seeds to "terminator seeds" and now to synthetic biology. The tools might change, but the quest to control life and society does not.

What I have called the "monoculture of the mind" cuts across all generations of technologies to control the seed.

- While farmers breed for diversity, corporations breed for uniformity.
- While farmers breed for resilience, corporations breed for vulnerability.
- While farmers breed for taste, quality, and nutrition, industry breeds for industrial processing and long-distance transport in a globalized food system.

Monoculture of industrial crops and monocultures of industrial junk food reinforce each other, wasting the land, wasting food, and wasting our health.

The privileging of uniformity over diversity, of quantity over quality of nutrition, has degraded our diets and displaced the rich biodiversity of our food and crops. It is based on a false creation boundary that excludes both nature's and farmers' intelligence and creativity. It has created a legal boundary to disenfranchise farmers of their seed freedom and seed sovereignty, and impose unjust seed laws to establish corporate monopoly on seed. Whether it be breeders' rights imposed through UPOV 91 or patents on seed or seed laws that require compulsory registration and licensing, an arsenal of legal instruments is being invented and imposed undemocratically to criminalize farmers' seed breeding, seed saving, and seed sharing.

Every seed is an embodiment of millennia of nature's evolution and centuries of farmers' breeding. It is the distilled expression of the intelligence of the earth and the intelligence of farming communities. Farmers have bred seeds for diversity, resilience, taste, nutrition, health, and adaption to local agroecosystems. Industrial breeding treats nature's contributions and farmers' contributions as nothing.

Just as the jurisprudence of terra nullius defined the land as empty and allowed the takeover of territories by the European colonists, the jurisprudence of intellectual property rights related to life-forms is in fact a jurisprudence of bio nullius—life empty of intelligence. The earth is defined as dead matter, so it cannot create. And farmers have empty heads, so they cannot breed.

The TRIPS Agreement and the Ethical Dimension

The deeper level at which the seed emergency is undermining the very fabric of life is the ethical dimension of this issue. We are all members of the earth family, stewards in the web of life. Yet corporations that claim legal personhood are now claiming the role of creator. They have declared seed to be their "invention," hence their patented property. A patent is an exclusive right granted for an "invention" that allows the patent holder to exclude everyone else from making, selling, distributing, and using the patented product. With patents on seed, this implies that the farmers' right to save and share seed is now in effect defined as "theft," an "intellectual property crime."

The door to patents on seed and patents on life was opened by genetic engineering. By adding one new gene to the cell of a plant, corporations claimed they had invented and created the seed, the plant, and all future seeds, which have now become their property. In other words, GMO meant God, move over.

In defining seed as their creation and invention, corporations like Monsanto shaped the global intellectual property and patent laws so that they could prevent farmers from seed saving and sharing. This is how the TRIPs Agreement of the World Trade Organization was born. Article 27.3(b) of the TRIPs Agreement states: "Parties may exclude from patentability plants and animals other than micro-organisms, and essentially biological processes for the production of plants or animals other than non-biological and microbiological processes. However, parties shall provide for the protection of plant varieties either by patents or by an effective sui generis system or by any combination thereof." Again, this protection on plant varieties is precisely what prohibits the free exchange of seeds between farmers, threatening their subsistence and ability to save and exchange seeds.

The TRIPS clause on patents on life was due for a mandatory review in 1999. India in its submission had stated, "Clearly, there is a case for re-examining the need to grant patents on lifeforms anywhere in the world. Until such systems are in place, it may be advisable to . . . exclude patents on all lifeforms." The African group stated, "The African Group maintains its reservations about patenting any life-forms as explained on previous occasions by the Group and several other delegations. In this regard, the Group proposes that Article 27.3(b) be revised to prohibit patents on plants, animals, micro-organisms, essentially biological processes for the production of plants or animals, and non-biological and microbiological processes for the production of plants or animals. For plant varieties to be protected under the TRIPS Agreement, the protection must clearly, and not just implicitly or by way of exception, strike a good balance with the interests of the community as a whole and protect farmers' rights and traditional knowledge, and ensure the preservation of biological diversity."

This mandatory review has been subverted by governments within the WTO: this long overdue review must be taken up to reverse patents on life and patents on seed. Life-forms, plants, and seeds are all evolving, self-organized, sovereign beings. They have intrinsic worth, value, and standing. Owning life by claiming it to be a corporate invention is ethically and legally wrong. Patents on seeds are legally wrong because seeds are not an invention. Patents on seeds are ethically wrong because seeds are life-forms; they are our kin, members of our earth family.

The worldview of bio nullius—empty life—unleashes violence and injustice to the earth, to farmers, and to all citizens. The violence to the earth is rooted in the denial of the creativity and the rights of the earth as well as in the displacement of diversity.

Biopiracy

The violence to farmers is threefold. First, their contribution to breeding is erased and what farmers have coevolved with nature is patented as an innovation. We call this "biopiracy." Patents on life are the hijacking of biodiversity and indigenous knowledge; they are instruments of monopoly control over life itself. Patents on living resources and indigenous knowledge are an enclosure of the biological and intellectual commons. Life-forms have been redefined as "manufacture" and "machines," robbing life of its integrity and self-organization. Traditional knowledge is being pirated and patented, unleashing this new epidemic of biopiracy. To end this new epidemic and to save the sovereignty and rights of our farmers, it is required that our legal system recognize the rights of communities, their collective and cumulative innovation in breeding diversity, and not merely the rights of corporations.

Moreover, patents lead to royalty collection, which is simply extortion in the name of technology and improvement. If the first colonization based on terra nullius gave us landlords and "Zameendari" who pushed 2 million people to death during the Bengal Famine, the new bioimperialism based on bio nullius has given us life lords—the biotechnology/seed/chemical industry, which has pushed 260,000 Indian farmers to suicide.

In Brazil, farmers have been fighting against seed giant Monsanto, most recently filing a lawsuit hoping to sue the company for over €6 million on the grounds that the company has been unfairly collecting royalties from the farmers. The seeds Monsanto has been collecting royalties on are from what are known as "renewal" seed harvests, meaning that the seeds have been collected from the previous harvest, a practice used for centuries. But because these seeds are from Monsanto's genetically modified plants, the company is demanding that farmers pay. Not only are these royalties unfairly enforced, but they are also pushing farmers deeper into debt that they cannot pay back, leaving them floundering in their fields of failed genetically modified crops.

In addition, when the genetically engineered crops contaminate neighboring farmers' fields, the "polluter pay" principle is turned on its head and corporations use patents to establish the principle of "polluter gets paid." This is what happened in the case of Percy Schmeiser in Canada as well as to thousands of farmers in the United States.

Owning and controlling life through patents and intellectual property rights was always the primary objective. Genetic engineering was the gateway to patents. Now the corporations are taking patents on conventionally bred and farm-saved seeds.

During the first Green Revolution (1950s/1960s), farmers' breeding was neglected. During the second Green Revolution (1990s), the biotech industries pushed for seed totalitarianism. Farmers' breeding is being criminalized. In 2004, an attempt was made to introduce a seed law in India that would require the compulsory registration of farmers' varieties. A seed satyagraha was started, and the law has not yet passed . . . *satyagraha* (force of the truth) was Gandhi's word for not cooperating with unjust laws. Gandhi said, "As long as the superstition exists that unjust law must be obeyed, so long will slavery exist." We need to globalize noncooperation with unjust seed laws. This is at the core of the movement for seed freedom. The stories of seed freedom are stories of courageous and creative individuals and organizations who are challenging unjust laws.

Patents on seed are unjust and unjustified. A patent or any intellectual property right is a monopoly granted by society in exchange for benefits. But society has no benefit in toxic, nonrenewable seeds. We are losing biodiversity and cultural diversity, we are losing nutrition, taste, and quality in our food. Above all, we are losing our fundamental freedom to decide what seeds we will sow, how we will grow our food, and what we will eat. Seed as a common good has become a commodity of private seed companies that unless protected and put back in the hands of our farmers is at risk of being lost forever.

Resistance to unjust seed laws through the seed satyagraha is one aspect of seed freedom. Saving and sharing seeds is another aspect. That is why Navdanya has worked with local communities to reclaim seed diversity and seed as a commons by establishing more than one hundred community seed banks. Across the world, communities are saving and exchanging seeds in diverse ways appropriate to their context. They are creating and recreating freedom—for the seed, for seed keepers, and for all life and all people.

When we save seed, we also reclaim and rejuvenate knowledge—the knowledge of breeding and conservation, the knowledge of food and farming. Uniformity as a pseudo-scientific measure has been used to establish unjust intellectual property rights (IPR) monopolies on seed. And IPR monopolies reinforce monocultures. Once a company has patents on seeds, it pushes its patented crops on farmers in order to collect royalties. Humanity has been eating thousands upon thousands (eighty-five hundred) of plant species. Today we are being condemned to eat GM corn and soy in various forms. Four primary crops—corn, soy, canola, and cotton—have all been grown at the cost of other crops because they generate a royalty for every acre planted. For example, India had fifteen hundred different kinds of cotton; now 95 percent of the cotton planted is GMO Bt cotton, for which Monsanto collects royalties. Over 11 million hectares of land are used to cultivate cotton, of which 9.5 million hectares are used to grow Monsanto's genetically modified Bt variety. Corn is cultivated on over 7 million hectares of land, but of this area 2,850,000 hectares are used for a "high-yielding variety" corn. Soy now covers an area of approximately 9.95 million hectares, and canola approximately 6.36 million hectares. The mass shift toward the cultivation of these crops not only threatens the diversity of other crops but threatens the health and well-being of natural resources such as the soil, as this monoculture approach to farming drains the earth of its nutrients.

To break out of this viciousness of monocultures and monopolies, we need to create virtuous cycles of diversity and reclaim our biological and intellectual commons. Participatory breeding of open-source seeds and participatory framing of open-source rights are innovations that deepen seed freedom. Seed freedom has become an ecological, political, economical, and cultural imperative. If we do not act, or have a fragmented and weak response, species will irreversibly disappear. Agriculture and the food and cultural spectrum dependent on biodiversity will disappear. Small farmers will disappear, healthy food diversity will disappear, seed sovereignty will disappear, and food sovereignty will disappear.

By speaking and acting strongly in one voice in defense of seed freedom as the Global Citizens Alliance, we can put the obscenity, violence, injustice, and immorality of patents on seeds and life behind us. Similarly, in another period slavery was made a thing of the past. Just as today corporations find nothing wrong in owning life, slave owners found nothing wrong in owning other humans. Just as people back then questioned and challenged slavery, it is our ethical and ecological duty—and our right—to challenge patents on seeds. We have a duty to liberate the seed and our farmers. We have a duty to defend our freedom and protect open-source seeds as a commons. This Global Citizens Alliance report on seed freedom is a kernel/seed that we hope will multiply and reproduce until no seed, no farmer, no citizen is bonded, colonized, or enslaved.

11

Food and Water

Food and water are our most basic needs. Without water, food production is not possible. That is why drought and water scarcity translate into a decline of food production and an increase in hunger. Traditionally, food cultures evolved in response to the water possibilities surrounding them. Water-prudent crops emerged in water-scarce regions, and water-demanding ones evolved in water-rich regions.

In the wet territories of Asia, rice cultures evolved and paddy field irrigation dominated. In the arid and semiarid tracts across the world, wheat, barley, corn, sorghum, and millet emerged as staples. In high-altitude regions, pseudo-cereals such as buckwheat provided nutrition. In the Ethiopian highlands, teff became the staple of choice. In deserts, pastoral cultivation was the basis of the food economy. Yet these diverse crops and agricultural styles are overlooked as food monoculture becomes the preferred method of production at the national, international, and corporate levels.

The water-use efficiency of crops is influenced by their genetic variation. Maize, sorghum, and millet convert water into biological matter most efficiently. Millet not only requires less water than rice, but it is also drought resistant, withstanding up to 75 percent soil moisture depletion. The roots of pulses and legumes allow efficient soil moisture utilization.

Since the Green Revolution, crops that produce higher nutrition per unit of water used have been called inferior and have been displaced by water-intensive crops. Water productivity has been ignored, the focus shifting to labor productivity. The replacement crops have produced not only unimpressive yields but low organic matter, reducing the moisture-conservation capacity of the soil.

Crop breeding in traditional societies took place keeping in mind the effect of droughts. In a participatory breeding experiment with farmers in the desert region of Rajasthan, India, the International Center of Research in Crops for the Semiarid Tropics (ICRISAT) discovered that the farmers preferred their indigenous varieties of millet, citing the crops' resistance to drought. The farmers also chose their varieties because of higher biomass yield in the form of straw, manure, and animal feed. The modern industrial plant breeding had bred out the drought-resisting traits of crops.[1]

Industrial Agriculture and Water Crisis

Industrial agriculture has pushed food production to use methods in which the water retention of soil is reduced and the demand for water is increased. By failing to recognize water as a limiting factor in food production, industrial agriculture has promoted waste. The shift from organic fertilizers to chemical fertilizers and the substitution of water-prudent crops with water-thirsty ones have been recipes for water famines, desertification, waterlogging, and salinization.

Droughts can be aggravated by climate change and soil moisture reduction. Drought caused by climate change—a phenomenon known as a meteorological drought—is linked to rainfall failure.[2] But even with normal rain, food production can suffer if the soil moisture retention has been eroded. In arid areas, where forests and farms are entirely dependent on the recharge of soil moisture, addition of organic matter is the only solution.[3] Soil-moisture drought occurs when organic matter necessary for moisture conservation is absent from soils. Prior to the Green Revolution, water conservation was an intrinsic part of indigenous agriculture. In the Deccan of south India, sorghum was intercropped with pulses and oilseeds to reduce evaporation. The Green Revolution replaced indigenous agriculture with monocultures, where dwarf varieties replaced tall ones, chemical fertilizers substituted organic ones, and irrigation displaced rain-fed cropping. As a result, soils were deprived of vital organic material, and soil moisture droughts became recurrent.

In drought-prone regions, ecologically sound agricultural systems are the only way to produce sustainable food. Three acres of sorghum use as much water as one acre under rice paddy cultivation. Both rice and sorghum yield forty-five hundred kilograms of cereals. For the same amount of water, sorghum provides 4.5 times more protein, 4 times more minerals, 7.5 times more calcium, and 5.6 times more iron, and can yield three times more food than rice.[4] Had agricultural development taken water conservation into account, millet would not have been called a marginal or inferior crop.

The advent of the Green Revolution pushed third-world agriculture toward wheat and rice production. The new crops demanded more water than millet and consumed three times more water than the indigenous varieties of wheat and rice.[5] The introduction of wheat and rice has also had social and ecological costs. Their dramatic increase in water use has led to the instability of regional water balances. Massive irrigation projects and water-intensive farming, by adding more water to an ecosystem than its natural drainage system can accommodate, have led to waterlogging, salinization, and desertification. Waterlogging occurs when the water table falls 1.5 to 2.1 meters. If water is added to a basin faster than it can drain out, the water table rises. About 25 percent of the irrigated land in the United States suffers from salinization and waterlogging.[6] In India, 10 million hectares of canal-irrigated land is waterlogged, and another 25 million hectares is under the threat of salinization.[7]

When waterlogging is recurrent, it is likely to lead to conflict between farmers and the state. In the Krishna basin, waterlogging at the Malaprabha irrigation project led to farmer rebellions. Before the introduction of the irrigation project, the semiarid land produced water-prudent crops such as jowar and pulses. The sudden climatic change, the intensive irrigation, and the cultivation of water-demanding cotton aggravated the problem. Intensive irrigation of black cotton soils, whose water-retention capacity is very high, quickly created wastelands. While irrigation has been viewed as a means to improve land productivity, in the Malaprabha area, it has had the opposite effect.[8] Farmers were shot by police when they refused to pay water taxes.[9] With the introduction of canal irrigation in the area, nearly 2,364 hectares of land have become waterlogged and saline.

Salinization is closely related to waterlogging. The salt poisoning of arable land has been an inevitable consequence of intensive irrigation in arid regions. Water-scarce locations contain large amounts of unleached soil;[10] pouring irrigation water into such soils brings the salts to the surface. When the water evaporates, saline residue remains. Today more than one-third of the world's irrigated land is salt polluted.[11] An estimated seventy thousand hectares of land in Punjab are salt affected and produce poor yields.[12]

The shift from rain-fed food crops to irrigated cash crops like cotton was expected to improve the prosperity of farmers. Instead it has led to debt.[13] Farmers borrowed money from banks for land development and for the purchase of seeds, chemical fertilizers, and pesticides. The total loans taken by the farmers increased from $104,449 in 1974 to more than $1.1 million by 1980. While farmers were struggling with unproductive land, banks were making payment demands. At the same time, irrigation authorities levied a development tax on water, known as a betterment levy. The latter increased from 38¢ to 63¢ per acre for jowar, and from 38¢ to over $1 per acre for cotton. A fixed tax of 20¢ per acre was effective with or without water use.[14]

In March 1980, the farmers formed the Malaprabha Niravari Pradesh Ryota Samvya Samithi (Coordination Committee of Farmers of the Malaprabha Ittihsyrf Area) and launched a noncooperation movement to stop paying taxes.[15] In retaliation, government authorities refused to issue the certificates needed by the farmers' children to enroll in schools. On June 19, 1980, the farmers went on a hunger strike in front of a local official's office. By June 30, ten thousand farmers had gathered to support those on hunger strike. A week later, a massive rally was held in Navalgund, and farmers went on another hunger strike.

When no response came from the authorities, the farmers organized a blockade. About six thousand farmers gathered in Navalgund, but their tractors were damaged and the rally was stoned by authorities. That same day, angry farmers seized the irrigation department, setting fire to a truck and fifteen jeeps. The police opened fire, killing a young boy on the spot. In the town of Naragund, the police opened fire at a procession of ten thousand people, shooting one youth. The protesters responded by beating a police officer and a constable to death. The protests rapidly spread to Ghataprabha, Tungabhadra, and other parts of Karnataka. Thousands of farmers were arrested, and forty were killed. In the end, the government ordered a moratorium on the collection of water taxes and the betterment levy.[16]

Unsustainable Agriculture: Water Waste and Destruction

The Aral Sea, the world's fourth-largest freshwater body, has been ruined by unsustainable agricultural activity. Rivers that recharge the lake are increasingly diverted toward the irrigation of 7.5 million hectares of cotton, fruit, vegetables, and rice.[17] Over the past few decades, two-thirds of the water has been drained away, salinity has gone up sixfold, and water levels have dropped by twenty meters. Between 1974 and 1986, the Syr Darya River never reached the Aral Sea; between 1974 and 1989, the Anu Darya failed to reach it five times. Instead, the water from these rivers feeds the Kara Kum irrigation canal near the Iranian border, eight hundred kilometers away.

In 1990, economist Vasily Selyunin commented of the Aral Sea: "The root of the problem is over irrigation, on a scale so vast that it has washed all the humus out of the soil. The loss had to be made good with shock doses of fertilizers. As a result, the earth has become like a junkie, unable to function without its fix." Fishing ports now lie forty to fifty kilometers from the Aral shores, and the fish catch has collapsed from twenty-five thousand tons a year to zero. Half of the population of the nearby city of Aralsk, Kazakhstan, has migrated. Unfortunately, as the Uzbek poet Muhammed Salikh points out, "You cannot fill the Aral with tears."[18]

Industrial farming is not just harming seas and rivers, but it is also impairing groundwater aquifers. The Ogallala Aquifer is irrigating farms in the High Plains of Texas. Each year, between 5 million and 6 million acre-feet of water are pumped from the Ogallala.[19] If the water continues to diminish at this rate, the only option left will be to shift to water-prudent, dryland farming or to abandon agriculture altogether. Sustainable agriculture policies would promote the former. Water markets promote the latter.

In the third world, fossil fuel–based mining technologies have devastated water resources. Energized groundwater pumping promulgated by the Green Revolution was considered efficient in terms of energy and horsepower use. An irrigation pump powered by a 7.5-kilogram electric motor took five hours and one person to irrigate an acre of wheat; in contrast, a Persian wheel requires up to sixty bullock hours and sixty human hours.[20] Whether the water withdrawal was inconsistent with groundwater recharge was not given any weight in the calculations of efficiency. Energized pumps that desiccated large areas of prime farmland in less than two decades were seen as more effective than the traditional methods such as the Persian wheel, which had sustainably supported agriculture for centuries.

Many of the solutions proposed to the problem of agricultural water waste deny water for food production altogether. Industrial shrimp farming is a case in point. The most obvious and important impacts of industrial aquaculture are land and water salinization and drinking water depletion. Paddy fields once fertile and productive are turning into what local people call graveyards. This is true not just in India. In Bangladesh, where shrimp farming is widespread, the amount of rice production has dropped considerably. In 1976, the country produced forty thousand metric tons of rice; by 1986, production had plummeted to thirty-six metric tons.[21] Thai farmers report similar losses, harvesting 150 sacks of rice per year instead of the 300 sacks they were harvesting before the introduction of shrimp farms to the region.[22]

Women have been particularly affected by the proliferation of the shrimp industry. Land has become a scarce commodity, and fights over patches of land are more and more frequent. Women in Pudukuppam, India, must walk one to two kilometers to fetch drinking water.[23] Wells have become sources of social tension. In the Indian village of Kuru, there is no drinking water available to the six hundred residents due to salinization. After the 1994 protests by the local women, water was supplied in tankers, with each household receiving only two pots per day for drinking, washing, and cleaning. "Our men need ten buckets of water to bathe after their fishing trips. What can we do with two pots?" is what women of coastal villages said to me that year. In Andhra Pradesh, the government supplied water by tankers from a distance of twenty kilometers for two years before it finally decided to move the five hundred families to another location. In a number of regions, relocation was not possible, and residents had no option but to use saline water for their crops and everyday needs.[24]

The United States is the most dramatic example of water waste in agriculture. In the western states, irrigation accounts for 90 percent of total water consumption. Irrigated land increased from 4 million acres in 1890 to nearly 60 million in 1977, of which 50 million were in the arid western states.[25] These areas are also affected by soil salinity because of salts dumped into rivers when irrigation waters drain. In a span of just thirty miles, the salt content of the Pecos River in New Mexico increases from 760 to 2,020 milligrams per liter.[26] In Texas, the salinity of the Rio Grande increases from 870 to 4,000 milligrams per liter in seventy-five miles.[27] Irrigation waters contribute five hundred thousand to seven hundred thousand tons of salt annually to the Colorado River: the loss of yield due to salt is estimated at $113 million a year.[28] In San Joaquin Valley, California, crop yields have declined by 10 percent since 1970, an estimated loss of $312 million annually.[29]

Water exhaustion is not the only problem caused by industrial agriculture. In Bengal, India, deep tube-well drilling has been identified as the cause of arsenic poisoning. In west Bengal, more than two hundred thousand people are dying or are permanently maimed due to arsenic poisoning.[30] In Bangladesh, 70 million people are poisoned by arsenic; in forty-three of Bangladesh's sixty-four districts, the arsenic level is around 0.05 milligram per liter, and in twenty districts, the level is above 0.5 milligram per liter; the permissible limit is 0.01 milligram per liter.[31] Many villages report arsenic of up to 2 milligrams per liter, two hundred times higher than the allowed level.

The Myth of Water Solution through Genetically Modified Crops

In 2001, I attended the World Economic Forum (WEF) in Davos, Switzerland, where, at a session on water, a representative from Nestlé suggested that genetic engineering would be a solution to water-intensive agriculture. He reasoned that genetic engineering could create drought-resistant crops that require little water. The obstacle, he argued, was the anti–genetic modification (GM) movement, which has prevented the introduction of drought-resistant varieties of GM crops.

The argument that genetic engineering will resolve the water crisis obscures two important points. First, peasants in drought-prone regions had bred thousands of drought-resistant crops, which were eventually displaced by the Green Revolution. Second, drought resistance is a complex, multigenetic trait, and genetic engineers have so far not been successful in engineering plants that possess it. In fact, the

GM crops currently in the field or in labs will aggravate the water crisis in agriculture. For instance, Monsanto's herbicide-resistant crops, such as its Roundup Ready soybeans or corn, have led to soil erosion. When all cover crops are killed by Monsanto's herbicide Roundup, rows of soy and corn leave soils exposed to tropical sun and rain.

Similarly, the heavily advertised vitamin A–rich golden rice increases water abuse in agriculture. Golden rice contains thirty micrograms of vitamin A per one hundred grams of rice. On the other hand, greens such as amaranth and coriander contain five hundred times more vitamin A, while using a fraction of the water needed by golden rice. In terms of water use, genetically engineered rice is fifteen hundred times less efficient in providing children with vitamin A, a necessary vitamin for blindness prevention. The golden rice promise is what I call "a blind approach to blindness prevention."

The myth of water solution by way of GM crops obscures the hidden cost of the biotech industry—the denial of fundamental rights of food and water to the poor. Investing in indigenous breeding knowledge and protecting the rights of local communities are more equitable and sustainable ways to ensure access to water and food to all.

Notes

1. *Participatory Breeding of Millets* (International Crops Research Institute for the Semi-arid Tropics, 1995).

2. Vandana Shiva et al., *Ecology and the Politics of Survival: Conflicts over Natural Resources in India* (New Delhi: Sage, 1991).

3. V. A. Kovda, *Land Aridization and Drought Control* (Boulder, CO: Westview, 1980); M. M. Peat and I. D. Teare, *Crop-Water Relations* (New York: Wiley, 1983).

4. Vandana Shiva, *The Violence of the Green Revolution: Third World Agriculture, Ecology, and Politics* (London: Zed Books, 1991), 70.

5. Ibid., 200.

6. Ibid.

7. Ibid.

8. Shiva et al., *Ecology and the Politics of Survival.*

9. Ibid.

10. Unleached soils contain salts that are not washed away by rain.

11. Shiva, *Violence of the Green Revolution,* 128.

12. Ibid., 129.

13. Vandana Shiva et al., *Seeds of Suicide* (New Delhi: Research Foundation for Science, Technology, and Ecology, 2001).

14. Shiva et al., *Ecology and the Politics of Survival,* 234.

15. Ibid., 235.

16. Ibid.

17. Robin Clarke, *Water: The International Crisis* (Cambridge, MA: MIT Press, 1993), 61.

18. William Ellis, "A Soviet Sea Lies Dying," *National Geographic,* February 1990.

19. Marq De Villiers, *Water: The Fate of Our Most Precious Resource* (New York: Houghton Mifflin, 2000), 44.

20. Shiva, *Violence of the Green Revolution,* 141.

21. Vandana Shiva and Gurpreet Karir, *Chemmeenkettu* (New Delhi: Research Foundation for Science, Technology, and Ecology, 1997).

22. Ibid.

23. Ibid.

24. Ibid.

25. Tim Palmer, *Endangered Rivers and the Conservation Movement* (Berkeley: University of California Press, 1986), 178.

26. Ibid., 192.

27. Mohamed T. El-Ashry, "Salinity Problems Related to Irrigated Agriculture in Arid Regions," *Proceedings of Third Conference on Egypt, Association of Egyptian-American Scholars* (1978): 55–75.

28. Mohamed T. El-Ashry, "Groundwater Salinity Problems Related to Irrigation in the Colorado River Basin and Ground Water," *Groundwater* 18, no. 1 (1980): 37–45.

29. De Villiers, *Water,* 143.

30. For further information on arsenic poisoning, visit the World Health Organization at www.who.int/water_sanitation_health/Arsenic/arsenic.htm.

31. For more reading on arsenic poisoning in Bangladesh, see Allan Smith, Elena Lingas, and Mahfuzar Rahman, "Contamination of Drinking-Water by Arsenic in Bangladesh: A Public Health Emergency," *Bulletin of the World Health Organization* 78, no. 9 (2000): 1093–1103, www.who.int/bulletin/pdf/2000/issue9/bu0751.pdf.

12

Soil, Not Oil

Securing Our Food in Times of Climate Crisis

Industrialized agriculture and globalized food systems have been put forth as sources of cheap and abundant food. However, food is no longer cheap. The era of cheap food and cheap oil is over. The food crisis, mainly triggered by rising prices, that emerged in 2007 and 2008 has led to food riots in many countries. From 2007 to 2008 the price of wheat increased by 130 percent.[1] The price of rice doubled during the first three months of 2008.[2] Biofuels, speculation, destruction of local food economies, and climate change have all contributed to the rise in food prices. Climate change is aggravated by industrialized, globalized agriculture based on fossil fuels, and the resulting climate crisis in turn impacts food security in numerous ways, including intensified floods such as those Iowa experienced in 2008 and intensified and extended droughts like the one Australia witnessed in 2007. Globalization has also led to the destruction of local food economies and increased control by corporations like Monsanto and Cargill over our food systems. Global integration of agriculture in effect means global control over the world's food supply.

In India, the World Bank–imposed structural adjustment program of 1991 and the WTO rules that came into force in 1995 have jointly worked to dismantle the public framework for food sovereignty and food security and to force the integration of India's food and agriculture systems with those of rich countries. This has resulted in a deep agrarian crisis and an emerging food crisis, with farmers' incomes crashing as food prices go through the roof. The food and agriculture crises are a direct result of policies of corporate globalization. Yet globalization is what the government is offering as a cure for globalization's ills.

Food prices started to rise as a result of connecting India's domestic market to global markets, especially the edible oil and wheat import markets. At first, in the early days of globalization, the agribusinesses that dominate trade lowered prices to grab markets. The dumping of soy in the 1990s is a prime example. Now that global corporations like Cargill have created import dependency, they are increasing prices. Additionally, speculation through futures trading is driving prices upward. Climate change and the diversion of foods to biofuels are also adding an upward pressure on international prices. The increase in international prices highlights the need to focus on food sovereignty. It makes both political and economic sense to focus on self-reliance in food and agriculture.

While millions go hungry, corporate profits have increased. Cargill saw profits increase by 30 percent in 2007; Monsanto's profits increased by 44 percent.[3] These profits will increase as corporate monopolies deepen. Monsanto increased the price of corn seed by $100 per bag to $300 per bag. For a thousand-acre farm in the United States, this means an increased cost of $40,000.[4]

The solution to the food crisis is to reclaim food sovereignty and rebuild local food economies based on ecological farming. This path also frees agriculture from its dependence on fossil fuels while increasing mitigation and adaptation to climate change. A shift from oil to soil addresses the triple crises of climate, energy, and food.

Eating Oil

Industrialized, globalized agriculture is a recipe for eating oil. Oil is used for the chemical fertilizers that go to pollute the soil and water. Oil is used to displace small farmers with giant tractors and combine harvesters. Oil is used to industrially process food. Oil is used for the plastic in packaging. And finally, more and more oil is used to transport food farther and farther away from where it is produced.

Fossil fuels are the heart of industrial agriculture. Fossil fuels are used to run the tractors and heavy machinery and to pump the irrigation water necessary for industrial farming. Industrial systems of food production use ten times more energy than ecological agriculture does, and ten times more energy than the energy in the food they produce.[5]

The *Stern Review: The Economics of Climate Change* has identified the following sources of greenhouse gas emissions responsible for climate change:

Greenhouse Gas Emissions, by Source:[6]

Power	24%
Industry	14%
Transport	14%
Buildings	8%
Land Use	18%
Agriculture	14%
Waste	3%
Other	5%

What the report does not mention is the particular kinds of agriculture, transport, and buildings that are responsible for the emissions. It fails to differentiate industrial, globalized agriculture, which is responsible for a large part of the 14 percent of emissions in agriculture, from nonindustrial, biodiverse ecological agriculture, which has much lower emissions and helps in carbon sequestration. It also does not break out the share of the 18 percent of emissions attributed to land use created when tropical forests are cut down to grow agricultural commodities, or the part of the 14 percent of transport emissions resulting from unnecessarily shipping and flying food around the world.

Localized, biodiverse, ecological agriculture can reduce greenhouse gas emissions by a significant amount while improving our natural capital of biodiversity, soil, and water; strengthening nature's economy; improving the security of farmers' livelihoods; improving the quality and nutrition of our food; and deepening freedom and democracy. Instead of focusing on achievable solutions, the Stern report promotes the pseudo-solution of carbon trading, which translates into business as usual for the agrochemical and agribusiness corporations profiting from globalized, industrialized agriculture.

An analysis of energy in the U.S. food chain found that on average, it takes ten calories of energy to produce one calorie of food. This is a net negative energy production system.[7] A shift to ecological, nonindustrial agriculture from industrial agriculture leads to a two- to sevenfold energy savings and a 5 to 15 percent global fossil fuel emissions offset through the sequestration of carbon in organically managed soil.[8] Up to four tons of CO_2 per hectare can be sequestered in organic soils each year.[9]

From field to table, the industrial, globalized food system is moving toward an increased dependence on fossil fuels. There have been dramatic changes in how food is produced, processed, and distributed over the last fifty years. The most significant changes include the following:

- The mechanization of agriculture and increased reliance on external inputs such as synthetic fertilizers, pesticides, feed, plastics, and energy.
- A major shift to highly processed and packaged food.
- The globalization of the food industry, characterized by an increase in food imports and exports. Of particular note is the rise in imports of fresh fruits and vegetables, with more produce sourced from farther afield.
- Supermarkets emerging as sales leaders, accompanied by the loss of small shops, markets, and wholesalers. Parallel to this trend is the concentration of supply into the hands of fewer, larger suppliers, partly to meet supermarkets' preferences for bulk, year-round supplies of uniform produce.
- Major changes in delivery patterns, with most goods now routed through supermarkets' regional distribution centers, and a trend toward the use of large heavy goods vehicles (HGVs) and just-in-time delivery, sometimes referred to as "warehouse on wheels."
- A switch from frequent food shopping on foot at small local shops to shopping by car at large out-of-town supermarkets.[10]

David Pimentel and Mario Giampietro have focused on the relationship between endosomatic and exosomatic energy. "Endosomatic energy is generated through the metabolic transformation of food energy into muscle energy in the human body. Exosomatic energy is generated by transforming energy outside of the human body by mechanical means, such as by burning oil in a tractor."[11] Pimentel and Giampietro found that it takes ten kilocalories of exosomatic energy to produce every one kilocalorie of food in the United States. The remaining nine kilocalories go to create waste and pollution, and increase entropy.[12] Part of this wasted energy is going into the atmosphere to contribute to climate change.

Industrial agriculture in the United States uses 380 times more energy per hectare to produce rice than does a traditional farm in the Philippines. And energy use per kilo of rice is 80 times more in the United States than in the Philippines. Energy use for corn production in the United States is 176 times more per hectare than on a traditional farm in Mexico and 33 times more per kilo.[13] One cow maintained and marketed in the industrial system requires six barrels of oil.[14] A 450-gram box of breakfast cereal provides only eleven hundred kilocalories of food energy but uses seven thousand kilocalories of energy for processing.[15]

Chemical industrial agriculture is based on the idea that soil fertility is manufactured in fertilizer factories. This was the idea that drove the Green Revolution, introduced in India in 1965 and 1966. In 1967, at a meeting in New Delhi, Norman Borlaug, the Nobel Prize–winning "father of the Green Revolution," was emphatic about the role of fertilizers in the new revolution. "If I were a member of your parliament," he told the politicians and diplomats in the audience, "I would leap from my seat every fifteen minutes and yell at the top of my voice, 'Fertilizers! . . . Give the farmers more fertilizers.' There is no more vital message in India than this. Fertilizers will give India more food."[16] Today, the Green Revolution has faded in Punjab. Yields are declining. The soil is depleted of nutrients. The water is polluted with nitrates and pesticides.

The fertilizer industry has now found Africa. The Rockefeller and Gates foundations have set up AGRA, the Alliance for a Green Revolution in Africa. However, AGRA will not be the site of a Taj Mahal for Africa's agriculture. The new Green Revolution for Africa is in fact the old Green Revolution for Asia. And as the Punjab experience shows, the Green Revolution was neither green in terms of ecological sustainability and conservation of the natural capital of soil-water-biodiversity nor revolutionary in terms of increasing equality and promoting justice for small and marginal peasants. This not-so-green revolution is now being proposed as a solution for hunger and poverty in Africa.

AGRA has a $150 million Program for Africa's Seeds Systems

(PASS) that seeks to transform farming in Africa. The strategy is based on promoting private seed companies and commercializing the seed supply, which AGRA assumes are necessary for improving Africa's farm productivity. It is also based on increasing the sale of chemical fertilizer. Gary Toenniessen of the Rockefeller Foundation writes in *Securing the Harvest,* "No matter what efficiencies genetic enhancement is able to build into crop plants, they will always draw their nutrition from external sources" and "No alternatives to the use of inorganic nitrogen currently exist for densely populated developing countries."[17] This ignores the successes in Asia, Africa, and Latin America of doubling and tripling farm productivity through biodiverse organic farming based on the farmers' breeding, biodiversity conservation, and agroecology. Not only are chemical fertilizers not necessary for farming; synthetic fertilizers actually harm the living processes in the soil that are responsible for soil fertility, plant growth, and production of healthy food.

Fertilizer advocates also ignore how the rising cost of oil affects fertilizer prices. Imported fertilizer costs from Rs. 55,000 to Rs. 60,000 per ton and is sold at Rs. 9,350 per ton. Rs. 45,000 per ton is paid through taxes collected to cover the subsidies. In India the shift to chemical agriculture has created the need for 4 to 4.8 million tons of synthetic diammonium phosphate (DAP). As only around 2 million tons are produced in India, the rest must be imported.

Fertilizer protests are taking place in Karnataka, where a farmer was killed when police opened fire on hundreds of farmers waiting for fertilizers. This was an entirely unnecessary tragedy. Similar incidents have occurred in Amrati, Vidarbha, Latur, Marathwada, and Maharashtra. First the Green Revolution made Indian farmers addicted to chemical fertilizer. Now globalization is making them dependent on imports.

While the soil and farmers die, agribusiness corporations like Cargill are making a killing. Cargill's fertilizer profits doubled from 2006 to 2007, with India paying 130 percent more for fertilizers and China 227 percent more during that period.[18]

Baron Justin von Liebig, a German chemist, carried out research in the latter part of the nineteenth century on the elements and chemicals required by plants for growth. He determined that the principal ingredients for soil fertility were nitrogen (N), phosphorus (P), and potassium (K). This is how the N-P-K mentality was born.

In 1909, Fritz Haber invented ammonium sulfate, a nitrogen fertilizer made by using coal or natural gas to heat nitrogen and hydrogen. The manufacture of synthetic fertilizers is highly energy intensive. One kilogram of nitrogen fertilizer requires the energy equivalent of two liters of diesel. One kilogram of phosphate fertilizer requires half a liter of diesel. Energy consumed during fertilizer manufacture was equivalent to 191 billion liters of diesel in 2000 and is projected to rise to 277 billion in 2030.[19]

Plants, however, need more than N-P-K. And when only N-P-K is applied as synthetic fertilizers, soils and plants, and consequently humans, develop deficiencies of trace elements and micronutrients. A pioneer of organic agriculture, Sir Albert Howard, explained: "A soil teeming with healthy life in the shape of abundant microflora and microfauna, will bear healthy plants, and these, when consumed by animals and man, will confer health on animals and man. But an infertile soil, that is, one lacking sufficient microbial, fungous, and other life, will pass on some form of deficiency to the plant, and such plant, in turn, will pass on some form of deficiency to animals and man."[20]

The millions of organisms found in soil are the source of its fertility. The greatest biomass in soil consists of microorganisms, fungi in particular. Soil microorganisms maintain soil structure, contribute to the biodegradation of dead plants and animals, and fix nitrogen. They are the key to soil fertility. Their destruction by chemicals threatens our survival and our food security. A Danish study analyzed a cubic meter of soil and found fifty thousand small earthworms, fifty thousand insects and mites, and 12 million roundworms. A gram of the soil contained thirty thousand protozoa, fifty thousand algae, four hundred thousand fungi, and billions of individual bacteria. It is this amazing biodiversity that maintains and rejuvenates soil fertility. To feed humanity we need to feed the soil and its millions of workers, including the earthworm.[21]

When I carried out research on the Green Revolution in Punjab, I found that after a few years of bumper harvests, crop failures at a large number of sites were reported, despite liberal applications of N-P-K fertilizers. The failure came from micronutrient deficiencies caused by the rapid and continuous removal of micronutrients by "high-yielding varieties." Plants quite evidently need more than N-P-K, and the voracious high-yielding varieties drew out micronutrients from the soil at a very rapid rate, creating deficiencies of such micronutrients as zinc, iron, copper, manganese, magnesium, molybdenum, and boron. With organic manure these deficiencies do not occur, because organic matter contains these trace elements, whereas chemical N-P-K does not. Zinc deficiency is the most widespread of all micronutrient deficiencies in Punjab. Over half of the 8,706 soil samples from Punjab exhibited zinc deficiency, which has reduced yields of rice, wheat, and corn by up to 3.9 tons, 1.98 tons, and 3.4 tons per hectare, respectively. Consumption of zinc sulfate in Punjab rose from zero in 1969–1970 to nearly 15,000 tons in 1984–1985 to make up for the artificially created zinc deficiency. Manganese is another micronutrient that has become deficient in Punjab soils. Sulfur deficiency, which was earlier noticed only in oilseed and pulse crops, has now been noticed in cereals like wheat.

The Green Revolution has also resulted in soil toxicity by introducing excess quantities of trace elements into the ecosystem. Fluorine toxicity from irrigation has developed in various regions of India. Twenty-six million hectares of India's lands are affected by aluminum toxicity. In the Hoshiarpur district of Punjab, boron, iron, molybdenum, and selenium toxicity has built up through Green Revolution practices and is posing a threat to crop production as well as animal health.

As a result of soil diseases and deficiencies, the increase in N-P-K application has not shown a corresponding increase in the output of rice and wheat. Wheat and rice yields have been fluctuating and even declining in most districts in Punjab, in spite of increasing levels of fertilizer use.

Experiments at the Punjab Agricultural University (PAU) are now beginning to show that chemical fertilizers cannot be substitutes for the organic fertility of the soil, and organic fertility can be maintained only by returning to the soil part of the organic matter that it produces. In the early 1950s, before the entry of the advisors of the Ford Foundation, when K. M. Munshi, India's agriculture minister at the time, referred to repairing the nutrient cycle, he was anticipating what agricultural scientists are today recommending for the diseased and dying fields of Punjab. And Howard's prediction, that "in the years to come, chemical manures will be considered as one of the greatest follies of the industrial epoch," is beginning to come true.[22]

Fertilizers block the soil capillaries that supply nutrients and water to plants. Infiltration of rain is stopped, runoff increases, and soil faces droughts, requiring ever more irrigation and ever more fossil fuels for pumping groundwater. Excess nitrogen in the root zone also denies nutrients to the plant. The negatively charged ions in the nitrates, the anions, take the cations, the positively charged ions of other elements, away from the root zone, thereby robbing the trees and plants of positive cations such as magnesium and calcium ions. Plants deficient in micronutrients create micronutrient deficiency in food and the human diet. And micronutrient deficiency leads to metabolic disorders.

Chemical fertilizers do not just destroy the soil and human health. They are also a major contributor to climate change because of pollution both from their production and from their use.

Long-distance globalized food systems, like the industrial food-production system they service, are contributing in a major way to greenhouse gas emissions. A study by the Danish Ministry of the Environment showed that one kilogram of food moving around the world generated ten kilograms of CO_2. "Food miles," which measure the distance food travels from where it is produced to where it is consumed, have increased dramatically as a result of globalization. As reported by environmental journalist Dale Allen Pfeiffer, "In 1981, food journeying across the US to the Chicago market traveled an average of 1,245 miles; by 1998, this had increased 22 percent, to 1,518 miles. In 1965, 787,000 combination trucks were registered in the United States, and these vehicles consumed 6,658 billion gallons of fuel. In 1997, there were 1,790,000 combination trucks that used 20.294 billion gallons of fuel. In 1979, David and Marcia Pimentel estimated that 60 percent of all food and related products in the US traveled by truck and the other 40 percent by rail. By 1996, almost 93 percent of fresh produce was moved by truck."[23]

A study in Canada has calculated that in 2003 food in Toronto traveled an average of 3,333 miles.[24] In the United Kingdom, the distance traveled by food increased 50 percent between 1978 and 1999.[25] A Swedish study found that the food miles of a typical breakfast would cover the circumference of the earth.[26]

The increase in food miles is related to fossil fuel and food subsidies, which allow food transported long distances to be cheaper than food produced locally. Thus, India imported 5.5 million tons of wheat in 2006, based on the argument that it was cheaper to import wheat from Australia and the United States than to transport it from Punjab in the north to Kerala and Tamil Nadu in the south. We should be reducing food miles by eating biodiverse, local, and fresh foods rather than increasing carbon pollution through the spread of corporate industrial farming, nonlocal food supplies, and processed and packaged food. We need to reduce CO_2 emissions by moving toward economic localization and satisfying our needs with the lowest carbon footprint. Economic globalization, on the other hand, only serves to increase CO_2 emissions. This total disconnect between ecology and economics is threatening to bring down our *oikos*, our home on this planet.

Imports, which add unnecessary food miles, are a direct result of free trade agreements. Transport accounts for one-eighth of oil consumption, and a large part of it goes for food. Take, for example, the wheat imports that resulted from the U.S.-India Knowledge Initiative on Agriculture. India is the second-largest producer of wheat in the world. Today, because of manipulation by the U.S. and Indian governments, it has suddenly emerged as a big importer of wheat. At the start of 2006 India's domestic production of wheat was projected to exceed domestic demand; it had been six years since India had needed to import wheat. However, because the country opened its domestic market to private corporations, foreign companies were able to buy so much wheat that the government found itself announcing that it would need to import wheat, initially purchasing 0.8 million tons from the Australian Wheat Board, the only company able to meet India's import standards. The company had previously been implicated in the Volcker report for giving Saddam Hussein's regime a $300 million kickback through Iraq's Oil-for-Food program.

As the year progressed and the Indian government continued to refuse a fair price for domestic wheat, it found itself once again forced to import wheat. This time it increased the price it was willing to pay and relaxed its import guidelines—allowing higher levels of toxins and pesticides. This meant that the big U.S. agribusinesses, primarily Cargill and ADM, could sell their wheat to India. India imported another 2.2 million tons, corporate agriculture gained, and food security suffered.[27]

From Food First to Export First

Until recently food has primarily been produced locally. Local food systems have evolved in accordance with local climates and biodiversity, which in turn have shaped the rich cultural diversity of food. We need both the diversity and the decentralization of local food systems to mitigate as well as adapt to climate change. However, both the World Bank and the World Trade Organization are forcing countries to dismantle their local food economies, export what they produce, and import what they need. The rise of "cash crop for export" policies are a result of World Bank structural adjustment policies. And the creation of import dependency is a result of World Bank conditionalities and WTO rules.

Sustainable agriculture is based on the sustainable use of natural resources—land, water, and agricultural biodiversity, including plants and animals. The sustainable use of these resources in turn requires that they are owned and controlled by decentralized agricultural communities, to generate their livelihoods and provide food. These three dimensions—ecological security, livelihood security, and food security—are essential elements of sustainable and equitable agriculture policy.

The current process of globalization of agriculture threatens to undermine all three of these dimensions. It is undermining ecological security by removing all limits on concentration of ownership of natural resources—land, water, and biodiversity—and encouraging nonsustainable resource exploitation for short-term profits. Trade liberalization of agriculture is not guided by the need to provide livelihood security for the two-thirds of India's people who are farmers or to provide food security for the poorer half of Indians and for India as a whole. Just the opposite: it severely threatens food security at the household, regional, and national levels.

The diversion of our natural resources from ecological maintenance, protection of livelihoods, and satisfaction of basic needs to luxury exports and corporate profits has been made possible because of the past three decades of agriculture policy. In that time agriculture has been made a state monopoly and run on massive debts and subsidies, while all ecological imperatives of sustainability have been ignored.

However, the new trade liberalization and globalization policies are not reducing the centralized control of agriculture; they are increasing it. Part of the reason people are not recognizing this new concentration and are misconceiving trade liberalization as a new freedom for farmers is because of the power shift from the nation-state to transnational corporations (TNCs). People have learned to recognize the lack of freedom built into the rule of the nation-state. They have not yet learned to recognize the lack of freedom intrinsic to corporate rule. As the state withdraws from agriculture, it is not returning power to farming communities and autonomous producers. It is instead facilitating the transfer of control over natural resources, production systems, markets, and trade to global agribusiness, further disempowering and dispossessing small farmers and landless laborers.

The WTO and the World Bank are pushing countries like India to move from food-first to export-first policies. A nutritional apartheid is thus being created, with the scarce land and water resources of the South being used for growing fruits and vegetables for the rich North and the elites of the South, and leaving the people of the South dependent on imports of food staples such as wheat, rice, and corn. Both sides of the equation add food miles to our daily bread. And while the destruction of local food systems and the dependence on globalized food supply are made to look "natural," they are the deliberate results of policy designed and driven by global agribusiness and supermarket chains. The step-by-step dismantling of India's local food markets exemplifies just how artificial and violent the globalization of food systems really is.

Failure of "Export-First" Policies

It is a sad irony that the creation of agricultural export zones (AEZs) intended to increase farm exports proved of no help to the vegetable grower. A bumper crop of potato did not bring farmers any profit; in fact, it ruined them, driving many to commit suicide. Despite the fact that the Indian government created three AEZs for potato cultivation, the potato could neither be exported nor utilized in the food-processing industry. Rather than increasing the exports of vegetables, the creation of AEZs has facilitated the import of vegetables.

- India is now the fifth-largest importer of raw vegetables, after the United States, the European Union, Japan, and Canada.
- The import bill for vegetables rose almost 20 percent in 2002; exports have been virtually static.
- India is spending three times more buying raw vegetables from world markets than it is earning from exporting them. The bill came to a huge $678 million in 2002. That was higher than the combined imports of Russia, Hong Kong, and Brazil. In contrast, India sold only $246 million worth of vegetables in 2002.[28]
- Exports of processed vegetables, fruits, and nuts plummeted from $70 million in 2001 to just $58 million in 2002.[29]

Experts have expressed fears that the large-scale diversion of land, capital, and other resources for crops like vegetables, flowers, and gherkins will severely affect food security. The very profitability of the cultivation of these crops needs to be properly assessed, taking fully into account the investment, the incentives given, and the value of the land and other forms of scarce natural resources diverted, or to be diverted. Since fruits and vegetables are perishable, they need to be transported in refrigerated trucks and by air. Trade in perishables is adding to the global carbon footprint.

The biodiverse, water-prudent, and drought-resilient agriculture of the South is being destroyed precisely when diverse and decentralized systems need to be conserved to reduce the impact of and increase resilience to climate change. On the one hand, drought is increasing as a result of climate change. On the other hand, it is increasing due to globalization of the food supply and diversion of land and water to produce cheap food for the rich in the North. Peasants and pastoralists are pushed off the land and denied access to water as corporate farming for exports takes over.

A fifty-gram bag of salad in the United Kingdom costs about £1 but wastes almost fifty liters of water. A mixed salad takes three hundred liters.[30] As Bruce Lankford of the University of East Anglia has stated, "We are exporting drought." Global retail chains like Tesco, Sainsbury's, and Walmart are increasingly sourcing fruits and vegetables from Africa and India. This is leading to the large-scale uprooting and impoverishment of farmers and is contributing to drought and desertification while increasing food miles and undermining food security and food sovereignty. While India is being made to grow vegetables for Europe, we are also being forced to import pesticide-laden wheat in spite of sufficient domestic production, which is further threatening farmers' livelihoods.

The poor are paying three times over—through increased vulnerability to climate change, through increased water scarcity as scarce water is used for export crops, and through the uprooting of communities from their land, villages, and homes to make way for wasteful globalized trade. Globalized trade in food is hurting the poor and the planet. It is putting the future of our food at risk for the short-term profits of global agribusinesses.

Soil, Not Oil: Making a Transition to Biodiverse, Organic, Local Food Systems

The industrialized, globalized food system is based on oil. It is under threat because of the inevitability of "peak oil." It is also under threat because it is more vulnerable than traditional agriculture to climate change, to which it has contributed. Industrial agriculture is based on monocultures. Monocultures are highly vulnerable to changes in climate as well as to diseases and pests.

In 1970 and 1971, America's vast corn belt was attacked by a mysterious disease, later identified as "race T" of the fungus *Helminthosporium maydis,* causing the southern corn leaf blight, as the epidemic was called. It left ravaged cornfields with withered plants, broken stalks, and malformed or completely rotten cobs. The strength and speed of the blight was a result of the uniformity of the hybrid corn, most of which had been derived from a single Texas male sterile line. The genetic makeup of the new hybrid corn, which was responsible for its rapid and large-scale breeding by seed companies, was also responsible for its vulnerability to disease. At least 80 percent of the hybrid corn in America in 1970 contained the Texas male sterile cytoplasm. As a University of Iowa pathologist wrote, "Such an extensive, homogenous acreage is like a tinder-dry prairie waiting for a spark to ignite it."[31]

Industrial agriculture is dependent on chemical fertilizers. Chemically fertilized soils are low in organic matter. Organic matter helps conserve the soil and soil moisture, providing insurance against drought. Soils lacking organic matter are more vulnerable to drought and to climate change. Industrial agriculture is also more dependent on intensive irrigation. Since climate change is leading to the melting of glaciers that feed rivers and in many regions of the world causing the decline in precipitation and increased intensity of drought, the vulnerability of industrial agriculture will only increase. Finally, since the globalized food system is based on long-distance supply chains, it is vulnerable to breakdown in the context of extreme events of flooding, cyclones, and hurricanes. While aggravating climate change, fossil fuel–dependent industrialized, globalized agriculture is least able to adapt to the change.

We need an alternative. Biodiverse, organic farms and localized food systems offer us security in times of climate insecurity while producing more and better food and creating more livelihoods.

The industrialized, globalized food system is based on oil; biodiverse, organic, and local food systems are based on living soil. The industrialized system is based on creating waste and pollution; a living agriculture is based on no waste. The industrialized system is based on monocultures; sustainable systems are based on diversity.

Living Soil

Every step in building a living agriculture sustained by a living soil is a step toward both mitigating and adapting to climate change. Over the past twenty years, I have built Navdanya, India's biodiversity and organic farming movement. We are increasingly realizing there is a convergence between the objectives of conserving biodiversity, reducing climate change impact, and alleviating poverty. Biodiverse, local, organic systems reduce water use and risks of crop failure due to climate change. Increasing the biodiversity of farming systems can reduce vulnerability to drought. Millet, which is far more nutritious than rice and wheat, uses only two hundred to three hundred millimeters of water, compared with the twenty-five hundred millimeters needed for Green Revolution rice farming. India could grow four times the amount of food it does now if it were to cultivate millet more widely. However, global trade is pushing agriculture toward GM monocultures of corn, soy, canola, and cotton, worsening the climate crisis.

Biodiversity offers resilience to recover from climate disasters. After the Orissa supercyclone of 1998 and the tsunami of 2004, Navdanya distributed seeds of saline-resistant rice varieties as "Seeds of Hope" to rejuvenate agriculture in lands that were salinated as a result of flooding from the sea. We are now creating seed banks of drought-resistant, flood-resistant, and saline-resistant seed varieties to respond to such extreme climate events. Climate chaos creates uncertainty. Diversity offers a cushion against both climate extremes and climate uncertainty. We need to move from the myopic obsession with monocultures and centralization to diversity and decentralization.

Diversity and decentralization are the dual principles needed to build economies beyond oil and to deal with the climate vulnerability that is the legacy of the age of oil. In addition to reducing vulnerability and increasing resilience, biodiverse organic farming also produces more food and higher incomes. As David Pimentel has pointed out, "Organic farming approaches for maize and beans in the US not only use an average of 30% less fossil energy but also conserve more water in the soil, induce less erosion, maintain soil quality, and conserve more biological resources than conventional farming does."[32]

After Hurricane Mitch struck Central America in 1998, farmers who practiced biodiverse organic farming found they had suffered less damage than those who practiced chemical agriculture. The ecologically farmed plots had on average more topsoil, greater soil moisture, and less erosion, and the farmers experienced less severe economic losses.[33]

Fossil fuel–based industrial agriculture moves carbon from the soil to the atmosphere. Ecological agriculture takes carbon from the atmosphere and puts it back in the soil. If ten thousand medium-sized U.S. farms converted to organic farming, the emissions reduction would be equivalent to removing over 1 million cars from the road. If all U.S. croplands became organic, it would increase soil-carbon storage by 367 million tons and would cut nitrogen oxide emissions dramatically.[34] Organic agriculture contributes directly and indirectly to reducing CO_2 emissions and mitigating the negative consequences of climate change.

Navdanya's work over the past twenty years has shown that we can grow more food and provide higher incomes to farmers without destroying the environment and killing peasants. We *can* lower the costs of production while increasing output. We have done this successfully on thousands of farms and have created a fair, just, and sustainable economy. The epidemic of farmer suicides in India is concentrated in regions where chemical intensification has increased costs of production. Farmers in these regions have become dependent on nonrenewable seeds, and monoculture cash crops are facing a decline in prices due to globalization. This is affecting farmers' incomes, leading to debt and suicides. High costs of production are the most significant reason for rural indebtedness.[35]

Biodiverse organic farming creates a debt-free, suicide-free, productive alternative to industrialized corporate agriculture and brings about a number of benefits. It leads to increased farm productivity and farm incomes while lowering costs of production. Pesticide-free and chemical-free production and processing bring safe and healthy food to consumers. We must protect the environment, farmers' livelihoods, public health, and people's right to food.

We do not need to go the Monsanto way. We can go the Navdanya way. We do not need to end up in food dictatorship and food slavery. We can create our food freedom. Biodiverse, organic, and local food systems help mitigate climate change by lowering greenhouse gas emissions and increasing absorption of CO_2 by plants and by the soil.

Organic farming is based on the recycling of organic matter; industrial agriculture is based on chemical fertilizers that emit nitrous oxides. Industrial agriculture dispossesses small farmers and converts small farms to large holdings that need mechanization, which further contributes to CO_2 emissions. Small, biodiverse, organic farms, especially in third-world countries, can be totally fossil fuel free. The energy for farming operations comes from animals. Soil fertility is built by recycling organic matter to feed soil organisms. This reduces greenhouse gas emissions. Biodiverse systems are also more resilient to droughts and floods because they have a higher water-holding capacity, making them more adaptable to the effects of climate change. Navdanya's study on climate change and organic farming has indicated that organic farming increases carbon absorption by up to 55 percent and water-holding capacity by 10 percent.

The environmental advantages of small-scale, biodiverse organic farms do not come at the expense of food security. Biodiverse organic farms produce more food and higher incomes than do industrial mono-cultures. Mitigating climate change, conserving biodiversity, and increasing food security go hand in hand.

The conventional measures of productivity focus on labor as the major input (and the direct labor on the farm at that) and externalize many energy and resource inputs. This biased productivity pushes farmers off the land and replaces them with chemicals and machines, which in turn contribute to greenhouse gases and climate change. Further, industrial agriculture focuses on producing a single crop that can be globally traded as a commodity. The focus on "yield" of individual commodities creates what I have called a "monoculture of the mind." The promotion of so-called high-yielding varieties leads to the displacement of biodiversity. It also destroys the ecological functions of biodiversity. The loss of diverse outputs is never taken into account by the one-dimensional calculus of productivity.

When the benefits of biodiversity are taken into account, biodiverse systems have higher output than monocultures. And organic farming is more beneficial for the farmers and the earth than chemical farming. When agroforestry is included in farming systems, carbon absorption and carbon return increase dramatically. Date palm and neem increase the carbon density in the soil by 175 and 185 percent, respectively.

Studies carried out by the USDA's National Agroforestry Center suggest that soil carbon can be increased by 6.6 tons per hectare per year over a fifteen-year rotation and wood by 12.22 tons per hectare per year. Since both soil and biomass sequester carbon, this amounts to removing 18.87 tons of carbon per hectare per year from the atmosphere.[36]

Soil and vegetation are our biggest carbon sinks. Industrial agriculture destroys both. By disrupting the cycle of returning organic matter to the soil, chemical agriculture depletes the soil carbon. Mechanization forces the cutting down of trees and hedgerows.

Organic manure is food for the community of living beings that depend on the soil. The alternatives to chemical fertilizers are many: green manures such as sesbania aculeata (dhencha), gliricidia, and sun hemp; legume crops such as pulses, which fix nitrogen through legume-rhizobium symbiosis; earthworms; cow dung; and composts. Farmyard manure encourages the buildup of earthworms by increasing their food supply. Soils treated with farmyard manure have from two to two and a half times as many earthworms as untreated soils. Earthworms contribute to soil fertility by maintaining soil structure, aeration, and drainage. They break down organic matter and incorporate it into the soil. The work of earthworms in soil formation was Darwin's major concern in his later years. Of worms he wrote, "It may be doubted whether there are many other animals which have played so important a part in the history of creatures."[37] The little earthworm working invisibly in the soil is the tractor, the fertilizer factory, and the dam combined. Worm-worked soils are more water stable than unworked soils, and worm-inhabited soils have considerably more organic carbon and nitrogen than the original soil. Their continuous movement forms channels that help in soil aeration. It is estimated that they increase the air volume of soil by up to 30 percent. Soils with earthworms drain four to ten times faster than those without, and their water-holding capacity is higher by 20 percent. Earthworm castings, which can amount to four to thirty-six tons per acre per year, contain five times more nitrogen, seven times more phosphorus, three times more exchangeable magnesium, eleven times more potash, and one and a half times more calcium than soil.[38] Their work on the soil promotes the microbial activity essential to the fertility of most soils.[39]

At the Navdanya farm in Doon Valley, we have been feeding the soil organisms. They in turn feed us. We have been building soil and rejuvenating its life. The clay component on our farm is 41 percent higher than those of neighboring chemical farms, which indicates a higher water-holding capacity. There is 124 percent more organic matter content in the soil on our farm than in soil samples from chemical farms. The nitrogen concentration is 85 percent higher, the phosphorus content 10 percent higher, and the available potassium 25 percent higher. Our farm is also much richer in soil organisms such as mycorrhiza, which are fungi that bring nutrients to plants. Mycorrhizal association makes food material from the soil available to the plant. Our crops have no diseases, our soils are resilient to drought, and our food is delicious, as any visitors to our farm can vouch. Our farm is fossil fuel free. Oxen plow the land and fertilize it.

By banning fossil fuels on our farm we have gained real energy—the energy of the mycorrhiza and the earthworm, of the plants and animals, all nourished by the energy of the sun.

Biodiversity: Our Natural Capital, Our Ecological Insurance

Biodiversity is our real insurance in times of climate change. Traditionally, farmers have increased their resilience by growing more than one crop. Sir Albert Howard saw in "mixtures," or biodiversity, the secret of sustainability and stability of farming in India. As he wrote in the 1940 classic on organic farming:

> Mixed crops are the rule. In this respect the cultivators of the Orient have followed Nature's method as seen in the primeval forest. Mixed cropping is perhaps most universal when the cereal crop is the main constituent. Crops like millets, wheat, barley, and maize are mixed with an appropriate subsidiary pulse, sometimes a species that ripens much later than the cereal. The pigeon pea, perhaps the most important leguminous crop of the Gangetic alluvium, is grown either with millets or with maize. The mixing of cereals and pulses appears to help both crops. When the two grow together, the character of the growth improves. Do the roots of these crops excrete materials useful to each other? Is the mycorrhizal association found in the roots of these tropical legumes and cereals the agent involved in this excretion? Science at the moment is unable to answer these questions: she is only now beginning to investigate them. Here we have another instance where the peasants of the East have anticipated and acted upon the solution of one of the problems which Western science is only just beginning to recognize. Whatever may be the reason why crops thrive best when associated in suitable combinations, the fact remains that mixtures generally give better results than monoculture.[40]

At Navdanya we have built on this ancient, time-tested knowledge, farming in nature's ways, based on biodiversity. Not only are we protecting biodiversity, we are increasing food production, farmers' incomes, and resilience in the face of climate change. On our farm, we have fields of seven crops (*saptarshi*), nine crops (*navdanya*), and twelve crops (*baranaja*). *Navdanya* in fact means "nine seeds" or "nine crops." Biodiverse fields always perform better than monocultures. They survive frost and drought, early rain and late rain, too much rain and too little rain. The baranaja (twelve crops) of bajra (pearl millet), maize (corn), safed chemi (beans), ogal (buckwheat), mandua (finger millet), jhangora (barnyard millet), urad (black gram), navrangi (rice bean), two varieties of koni (horsetail millet), lobia (bean), and til (sesame) produced more food and earned more than twice that of the corn monoculture. The baranaja or navdanya system of farming is a guarantee against hunger and an insurance against crop failure due to climate variability. In diverse parts of the country, biodiverse agricultural systems outperform monocultures.

Symbiosis among plants contributes to an overall increase in productivity of the crops. In the Western Ghats, a small farm typically has 1.5 acres of paddy, 0.5 acres of areca nut, and a kitchen garden with vegetables that include eggplants, beans, cucumbers, chilies, and small gourds. Likewise, in the eastern Himalaya, especially in Sikkim, the dominant land use is the sustainable *Alnus*-cardamom agroforestry system, in which cardamom plants and *Alnus* trees are intercropped to the benefit of cardamom production. In Rajasthan, too, in the arid tract of Jodhpur and parts of western Rajasthan, neem-based agroforestry and khejri (*Prosopis cineraria*), wherein crops like bajra, sorghum mung, moth bean, and corn are grown together, have fulfilled the nutritional requirements of the communities.[41]

A recent study conducted by Navdanya in four districts of West Bengal shows that multiple cropping (MC) is economically more efficient than modern intensive-chemical farming systems that cultivate monocultures. The net value of the annual production of an average MC farm is uniformly more than that of an average monoculture farm. The MC farms of East Medinipur district are sown with a wide range of crops, both in a sequential rotation and intercropped. Some of these farms—mostly smaller than a hectare in size—grow over fifty types of crops, excluding rice. The rain-fed farms of Bankura district are comparatively less diverse, hardly exceeding fourteen crops a year, including rice. The irrigated monoculture farms, by contrast, grow two rice varieties in Bankura district and three rice varieties in East Medinipur district (all high-yielding varieties, or HYV). The cost of all inputs (water for irrigation, seeds, agrochemicals, labor, and energy) was calculated to compare the relative gain in output value of the modern monoculture farms with that of the MC farms, and the remarkable finding was that the value of farm produce increases significantly with greater diversity of crops. Farmers explain this as "farm fatigue" from monoculture and intensive use of agrochemicals—an essential feature of modern agriculture.

These data contradict the prevailing mainstream agronomic view that intensive cultivation of a staple crop enhances productivity. A majority of farmers in Bankura and Medinipur have realized over years that the yield of monoculture farms is unsustainable. Many of these farmers have reverted back to traditional farming systems involving folk crop varieties. Some of them have experimented with a hybrid system of rotational cropping of a large number of "secondary" crops and an HYV rice. However, most of these MC farmers reported that the cost of the inputs eats away at the extra production of HYV rice and that the best means to cut down on the extraneous inputs is to "give the land a recess" by growing vegetables and fruits for a few years before replanting it with rice.[42]

Small biodiverse farms have higher productivity than monocultures, which are a necessary aspect of industrial agriculture based on external inputs. Higher biological productivity translates into higher incomes for small farmers. In Rajasthan, monocultures of pearl millet yielded Rs. 2,480 of net profit per acre, whereas a biodiverse farm of pearl millet, moth bean, and sesame yielded a net profit of Rs. 12,045 per acre, nearly five times the profit. In Uttarankhand, a monoculture of paddy yielded Rs. 6,720 per acre, whereas a biodiverse farm yielded Rs. 24,600 per acre, three and a half times the profit. In Sikkim, a monoculture farm of corn yielded Rs. 4,950 per acre while a mixed farm of corn, radish, lahi saag, and peas yielded Rs. 11,700 per acre. Navdanya's rice and wheat farmers have doubled production by using indigenous seeds and organic methods. Jhumba rice in Uttarankhand produces 176 quintals per hectare compared with 96 quintals per hectare of kasturi, a high-yielding rice variety. The paddy yields are 104 and 56 quintals per hectare, respectively. Farmers in West Uttar Pradesh have gotten yields of 62.5 quintals per hectare using a native wheat variety for organic production, compared with 50 quintals per hectare for chemically produced wheat.

Conservation of native seeds and biodiverse ecological farming have yielded incomes two to three times higher than monoculture farming, and eight to nine times higher than industrial systems using genetically engineered seeds.

Seeds of Freedom, Seeds of Life

Twenty-one years ago, in 1987, I started to save seeds to create a different future than the one envisioned by the biotechnology industry—in which all seeds are genetically engineered and patented. The vision for seed freedom evolved as Navdanya. *Navdanya* means "nine seeds," and it also means "the new gift." Through Navdanya, we have brought the new gift of ancient seeds to our farmers. Navdanya builds community seed banks based on rescuing, conserving, reproducing, multiplying, and distributing native varieties or farmers' varieties—varieties evolved and bred over millennia. On the one hand, our seed saving defends seeds as a commons—resisting through our daily actions the degraded, immoral, uncivilized idea that seeds are the "intellectual property" of corporations, and that saving them is a crime. On the other hand, Navdanya's seed banks are the basis of another food economy, one based on biodiversity and cultural diversity, on sustainability, and on the future.

The dominant food economy is based on monopolies and monocultures, on industrialization of production and globalization of distribution of a handful of crops—corn, soy, rice, and wheat. This economy has pushed 1 billion people into hunger, another 2 billion into obesity. It is killing species and farmers. One hundred fifty thousand small farmers of India have committed suicide because they were forced to buy costly, unreliable seed every year from corporations like Monsanto, which collect exorbitant royalties.

Navdanya's seed saving spreads seeds of life instead of seeds of death. We spread seeds of hope instead of seeds of hopelessness and despair. We spread seeds of freedom instead of seeds of slavery and seeds of suicide. After the 2004 tsunami, our salt-resistant rice varieties rebuilt the devastated agriculture of Tamil Nadu. Our seeds of *Dehradun* basmati gave us the strength to fight RiceTec of Texas, which had patented basmati rice. Our seeds of native wheat varieties inspired us to fight Monsanto when it patented low-gluten wheat. Our seeds teach us lessons in diversity and democracy. From our seeds we learn how to defend freedom of biodiversity and freedom of farmers in an age of corporate monopolies, terminator technologies, and globalized monocultures.

A false assumption is growing that we need genetic engineering to deal with climate change. It is false for a number of reasons. First, nature and farmers have evolved, and continue to evolve, varieties of plants that are resilient to drought, floods, and salinization due to cyclones, three major impacts of climate change. In Navdanya community seed banks, we have crops like millet that can withstand severe drought; we have rice that grows eighteen feet tall and can survive the floods of the Ganges basin. We have rice that can tolerate salt, which we distributed after the Orissa cyclone and the tsunami. The salt-tolerant varieties we have saved, multiplied, and distributed include *Kalambank, Kartikpatini, Chakaakhi, Dhala patini, dudeshwar, lilabati,* and *luna* (which means "salt"). Flood-resistant rice varieties include *Jalaj, Abhiman, Bhutna, Sada dhepa, Sada pankul, Jal kalas* (which means "the water pot"), *Bagada, Betana, Bhundi, Champi, Fareka, Indrijiba, Madia,* and *Kala bagada.* In regions that face floods and the ingress of saltwater from the sea, these varieties offer security in the face of climate change.

But rice does not grow only in wet regions. We have also saved hundreds of drought-tolerant rices, such as *Bhat kalon, Chaina, Gyarsu, Jhumka, Ramjawain ukhri, Asan leija, Bhut moni, Kaya, Loha, Gora, Nata,* and *Raja manik.* These are rain-fed rices that need no irrigation.[43] And there are many varieties of other crops that have the potential to evolve and help us face the growing water scarcity. The drought-resistant native wheats, and the millets like ragi, jhangora, koni, bajra, and jowar are "forgotten foods" that are the foods of the future.[44]

Genetic engineering will only allow corporations to take these seeds, appropriate their traits, patent them, and prohibit their use by farmers who don't make heavy royalty payments. Genetic engineering does not *create* the traits for drought, flood, and salt tolerance; it merely allows the *transfer* of traits across species.

In Navdanya we are creating community seed banks for climate emergencies so that the widest varieties of crops are available to communities to respond to climate-related disasters. And this diversity is available as a commons. Diversity and the commons are the two types of insurance we have in times of uncertainty and unpredictability. Diversity gives us the basis to evolve and adapt under changing conditions. Climate change is not a linear phenomenon that creates warming everywhere, or more rain or less rain. It is nonlinear, and it is better to talk of climate chaos than climate change or global warming. Our community seed banks of climate change–resilient varieties become even more important as the gene giants like Monsanto, DuPont, Syngenta, and Dow apply for patents on climate traits in crops such as drought tolerance and flood tolerance.[45]

In the context of climate chaos, diversity is the basis of adaptation. Monocultures and uniformity are recipes for breakdown. While at the ecological level, we need diversity to respond to climate chaos, at the social and political levels, we need the commons. Monopolies and concentration of ownership of resources enhance vulnerability in periods of chaos.

The mechanistic paradigms on which genetic engineering, intellectual property rights and patents on seeds, and globalized corporate control over food systems are based have given us climate chaos. They cannot help us adapt and evolve. As Einstein said, you cannot solve a problem using the mind-set that created it. Mechanistic thought creates monocultures of the mind. We must move beyond monocultures to protect the earth's rich diversity and use it to respond to climate chaos.

Humanity has eaten over eighty thousand edible plants over the course of its evolution. More than three thousand have been used consistently. However, we now rely on just eight crops to provide 75 percent of the world's food. Monocultures are destroying biodiversity, our health, and the quality and diversity of food. Monocultures have been promoted as an essential component of industrialization and the globalization of agriculture. They don't in fact produce more food. All they produce is more control and profits—for Monsanto, Cargill, and ADM. They create pseudo-surpluses and real scarcity by destroying biodiversity, local food systems, and food cultures.

In 1998, India's indigenous edible oils—made from mustard, coconut, sesame, linseed, and groundnut and processed in artisanal cold-press mills—were banned, with "food safety" used as an excuse. At the same time, restrictions on the import of soy oil were removed. The livelihoods of 10 million farmers were threatened. One million oil mills in villages were closed. More than twenty farmers were killed while protesting against the dumping of soy on the Indian market, which was leading to a fall in the price of domestic oilseed crops. Millions of tons of artificially cheap GM soybean oil continue to be dumped on India.

Women from the slums of Delhi formed a movement to reject soy and bring back mustard oil. "Sarson bachao, soyabean bhagao" (Save the mustard, drive away the soybean) was the women's call from the streets. We did succeed in bringing back mustard through our *satyagraha* (noncooperation with the ban).

The same companies that dumped soy on India—Cargill and ADM—are destroying the Amazon to grow soy. Millions of acres of rain forest—the lungs, the liver, the heart of the global climate system—are being burned to grow soy for exports. Armed gangs take over the forest and use slaves to cultivate soy. When people like Sister Dorothy Stang oppose the destruction of the forests and the violence against people, they are assassinated.[46]

While people in Brazil and India are being threatened directly by these agribusiness monocultures, people in the United States and Europe are also at risk. Eighty percent of soy production is being used as cattle feed to provide cheap meat. Cheap meat that is, in effect, destroying both the Amazon rain forest and people's health in rich countries. One billion people are without food because industrial monocultures robbed them of their livelihoods in agriculture and their food entitlements.[47] Another 1.7 billion are suffering from obesity and food-related diseases. Monocultures lead to malnutrition—for those who are underfed as well as those who are overfed.

Corporations are forcing us to eat untested GMO food. Soy is in 60 percent of all processed food. It has high levels of isoflavones and phytoestrogens, which produce hormone imbalances in humans. Traditional fermentation, as in the food cultures of China and Japan, reduces the levels of isoflavones. The promotion of soy in food is a huge experiment promoted with $13 billion in subsidies from the U.S. government between 1998 and 2004, and $80 million a year from the American soy industry.[48] Nature, culture, and people's health are all being destroyed. Local food cultures have rich and diverse alternatives to soy. For protein we have thousands of varieties of beans and grain legumes—the pigeon pea, chickpea, mung bean, urd bean, rice bean, adzuki bean, moth bean, cowpea, lentil, horse gram, and fava bean. For edible oils we have sesame, mustard, linseed, saffola, sunflower, groundnut.

With the spread of monocultures and the destruction of local farms, the food system has become dependent on fossil fuels—for synthetic fertilizers, for running giant machinery, for long-distance transport. We are increasingly eating oil, not food, threatening the planet and our health.

Moving beyond monocultures of the mind has become an imperative for repairing the food system. Biodiverse small farms have higher productivity and they generate higher incomes for farmers. And biodiverse diets provide more nutrition and better taste. Bringing back biodiversity goes hand in hand with bringing back small farms. Corporate control thrives on monocultures. Citizens' food freedom depends on biodiversity. Human freedom and the freedom of other species are mutually reinforcing, not mutually exclusive.

Rebuilding Local Food Communities

The globalized food system is causing destruction at every level. Biodiversity is being destroyed in favor of monocultures of corn, soy, and canola. Food has been reduced to a commodity. And the commodity can run a car, feed animals in factory farms, or feed people. Uniqueness, distinctiveness, quality, nutrition, and taste are no longer in the equation.

Farmers are being destroyed because prices of farm products are driven down through a combination of monopolistic buying by global corporations and dumping of subsidized products on the market. In the meantime, food prices keep rising for the poor, and hunger grows. The long-distance transport of food pollutes the atmosphere with carbon dioxide emissions from fossil fuels. No one is gaining from globalized trade in food except the corporations. Localization of food systems to reduce food miles is a climate-change imperative. It is also a food sovereignty and human rights imperative. Small farmers will survive only in the context of vibrant and robust local food economies.

Localization is also a food-security imperative. Short supply chains ensure better democracy in distribution, better-quality food, fresher food, and more cultural diversity. In India, the movement for retail democracy is a vital part of keeping local markets alive. Across the world, farmers' markets are reappearing. The search for local foods to reduce food miles and create more intimate food systems has created a new dichotomy between "organic" and "local." In my view this is a false dichotomy. To be organic means to be whole and wholesome—for the earth, for our bodies. Food that could have been grown next door but has been imported from thousands of miles away is not organic by any ecological standards. If we care about getting rid of toxins in our food, we should also care about the atmospheric pollution that is causing climate change. They are two facets of ecological destruction. A nonviolent, wholesome food system should have place for neither. Organic that leaves out food miles is not fully organic. Organic that leaves us feeling strangers on the land is not truly organic.

As Michael Pollan observes in his book *The Omnivore's Dilemma,*

> One of the key innovations of organic food was to allow some more information to pass along the food chain between the producer and the consumer—an implicit snatch of narrative along with the number. A certified organic label tells a little story about how a particular food was produced, giving the consumer a way to send a message back to the farmers that she values tomatoes produced without harmful pesticides or prefers to feed her children milk from cows that haven't been injected with growth hormones. The word organic has proved to be one of the most powerful words in the supermarket: Without any help from government, farmers and consumers working together in this way have built an $11 billion industry that is now the fastest growing sector of the food economy.
>
> Yet the organic label itself—like every other such label in the supermarket—is really just an imperfect substitute for direct observation of how a food is produced, a concession to the reality that most people in an industrial society haven't the time or the inclination to follow their food back to the farm, a farm which today is apt to be, on average, fifteen hundred miles away. So to bridge that space we rely on certifiers and label writers and, to a considerable extent, our imagination of what the farms that are producing our food really look like. The organic label may conjure an image of a simpler agriculture, but its very existence is an industrial artifact. The question is, what about the farms themselves? How well do they match the stories told about them?[49]

Organic farming is based on ecological processes and principles of agroecology. It is also based on human communities working in cooperation and with dignity and freedom.

There was an old conflict between chemical-industrial agriculture and organic farming. There is a new conflict emerging between *authentic organic,* based on small, biodiverse farms, and *pseudo-organic,* based on large-scale, monoculture corporate farms that grow for export. Authentic organic farming is based on biodiversity, small family farms, local markets, and fair trade. Organic farming emerged as a systemic alternative to industrial agriculture, which destroyed biodiversity, polluted ecosystems and food with agrochemicals, uprooted and displaced small farmers, and undermined local markets through subsidized long-distance transport.

Pseudo-organic farming destroys small farms and uproots small farming communities to create large export-oriented industrial farms in which farmers are viewed as laborers and serfs, instead of sovereign producers. Pseudo-organic farming is based on the destruction of biodiversity and creation of monocultures. It does not abide by the essential ecological processes of renewal of soil fertility, rejuvenation of water, and biodiversity. It merely substitutes chemical inputs with "organic" inputs. This is input substitution, not agroecology.

Agroecology is the scientific basis of authentic organic farming. Authentic organic practices are based on principles of self-organization—from the level of the organism to the farm and agroecosystem to the community. Ecologically, self-organization refers to the capacity of living organisms and agroecosystems to renew fertility by rejuvenating soil microorganisms and recycling organic matter; to manage pests through building resilience and maintaining a pest-predator balance; to conserve water; and to conserve and renew biodiversity. Seed giving rise to seed and earthworms rejuvenating soil fertility are examples of the self-organizing capacity of nature and living systems, which are the basis of a sustainable agriculture.

Socially, self-organization is encapsulated in Gandhi's *swaraj* (self-rule, self-governance, self-organization). It is the basis of food sovereignty—the right to produce in freedom. Social and ecological self-organization reinforce each other. Only small farmers working in cooperation with the soil and plants can provide the care and attention required to facilitate nature's self-organization. Food sovereignty, therefore, rests on agroecology. Both are built on the principle of self-organization. Self-organized production rests on the principles of agroecology, and self-organized distribution rests on the principles of localization—local consumption through local markets. Such economic self-organization ensures that local food needs are met and local food security and livelihoods strengthened, preventing malnutrition, hunger, poverty, and unemployment. It also provides the ground for cultural diversity in food systems, supported by biodiversity in agricultural systems.

Pseudo-organic agriculture is built on the destruction of the self-organizing capacity of human communities and agroecosystems. It mimics industrial agriculture, focusing on large-scale production for export, uprooting small farmers, and undermining people's food security and sovereignty. Large-scale, industrial-style, export-oriented pseudo-organic farms are run by giant corporations for profits at the expense of the health of the earth, diverse species, and local communities. The entry of multinational corporations in organic agriculture is based on land reforms for the rich, which usurp the lands of poor and marginal farmers. This is what is happening in Punjab, where the government is taking over land by force from small farmers and handing it over to corporations planning to export "organic" vegetables and fruits. Just as chemical farming and GM seeds are driving farmers into debt and suicide, pseudo-organic farming, which is corporate and export driven, is also killing farmers by taking away their land, their livelihoods.

An agriculture that destroys biodiversity, uproots local farmers, and leaves local communities without food is not worthy of the label "organic." To be organic is to be just and fair. An agriculture that turns rural areas into graveyards for farmers cannot be called organic. *Organic* means life giving. Authentic organic farming gives life. Pseudo-organic farming ends life. To remain authentic, organic farming must be biodiverse, it must stay in the hands of small farmers, and it must deepen food sovereignty.

In Navdanya, we work on the following principles of organic and local:

- *Food for the soil and its millions of microorganisms*
  Organic can be organic only if the food rights of millions of soil organisms are protected. This involves the law of return, of growing food for the soil, not just growing commodities for the market. In fact all "developments" in industrial agriculture are methods of increasing commodity production at the expense of the soil. The Green Revolution, with its chemical-intensive dwarf varieties, killed the soil organisms and used techniques that did not return organic matter to the soil. Genetically engineered herbicide-resistant crops, like Roundup Ready soy and corn, deliberately kill vegeta-

tion that would have gone back to feed the soil. Feeding markets while starving the soil is a recipe for hunger and desertification. If we feed the soil, we will also feed people, and even have quality production for the market.

- *Food and nutrition for the farming family*
The tragedy of industrialized, globalized agriculture is that while commodity markets grow, people starve. More than 1 billion people are now permanently hungry. Most of them are from rural areas. Many of them are food producers. They are denied food either because their soils have been desertified or because chemical agriculture and costly seeds have got them into debt or because they are growing cash crops like cotton and coffee, which bring insufficient returns because globalized trade has pushed down farm prices, or because they have been pushed off the land. It is criminal that our *annadatas,* our food providers, should themselves be hungry. That is why we ensure that every producer family that is a member of Navdanya first grows healthy and nutritious foods for the household and only trades any surplus.

- *Food for local communities*
Everyone must eat. If food is not grown locally, local communities will have to import their food from somewhere far away. That food will be more contaminated and adulterated and less safe. If local communities do not eat local produce, biodiversity will disappear from our farms and cultural diversity will disappear from our diets, making both the land and its people poorer.

- *Unique products for long-distance trade and exports*
Every part of the earth is productive. Every culture on the earth has evolved its diet according to the particular ecosystem it inhabits. As much as possible, food staples must be grown locally, both to produce what the ecosystem is best suited for and to produce what local cultures have adapted themselves to. Trade in food must be restricted to what cannot be grown locally; it must be restricted to foods with both a high value and a small ecological footprint in terms of land and water use.

Different vegetables and fruits grow in different climates. It is wrong to grow temperate-zone vegetables in the tropics and fly them back to rich consumers. This uproots local peasants, creates hunger and poverty, and destroys local agrobiodiversity. It also blocks the potential for localization in importing countries. Since vegetables and fruits are perishable, transporting them long distances is highly energy intensive, contributing to climate change. In India, the home of the mango, the Alphonso is traded and eaten only in Maharashtra and Goa, where it grows, and the Dasheri is largely eaten in the northern regions where it grows. Global trade in perishables destroys the biodiversity of fruits and vegetables. One kind of Chiquita banana, one kind of Washington apple ends up on every table. Local production for local consumption is the best way to conserve biodiversity, taste, and quality.

Spices are a perfect candidate for long-distance trade. Tiny quantities are needed to add flavor to food. Spices grow in very specific ecosystems. They cannot be grown everywhere. They give high value with low volumes. This benefits the producer, who can also grow food. In Karnataka, spice growers use 10 percent of their land for spice gardens of pepper, cardamom, and areca nut and 10 percent for paddy for local consumption. These gardens have existed for centuries and are a model for farming that supports trade but is not destroyed by trade. "Spice of life trade" is justified when it enriches the giver and the receiver.

Relocalization of our food systems has become an ecological and social imperative. Richard Heinberg, one of the preeminent theorists of peak oil, has pointed out that this will require the deindustrialization of agriculture. "The general outline of what I mean by de-industrialization is simple enough: this would imply a radical reduction of fossil fuel inputs to agriculture, accompanied by an increase in labour inputs, and a reduction in transport, with production being devoted primarily for local consumption. Fossil fuel depletion almost ensures that this *will* happen. But at the same time, it is fairly obvious that if we don't *plan for* de-industrialization, the result would be catastrophe."[50]

Rob Hopkins, the inspiration behind the new transition culture movement, elaborates on how energy-descent plans, or "powering down" of fossil fuel use, can be a "powering up" of the quality of life. "The essence of an energy descent plan is that it creates a vision of an abundant low energy future. While the transition away from fossil fuels will be a task of unprecedented proportions, at the same time it offers the potential for a society which is better in many ways, more connected to nature, healthier with more meaningful work, access to nutritious food, enhanced social capital, and more cooperation."[51]

Climate Change and the Two Carbon Economies: Biodiversity versus Fossil Fuels

Reductionism seems to have become the habit of the contemporary human mind. We are increasingly talking of climate change in the context of "the carbon economy." We refer to "zero carbon" and "no carbon" as if carbon exists only in fossilized form under the ground. We forget that the cellulose of plants is primarily carbon. Humus in the soil is mostly carbon. Vegetation in the forests is mostly carbon. It is living carbon. It is part of the cycle of life.

The problem is not carbon per se but our increasing use of fossil carbon that was formed over millions of years. Today the world burns four hundred years' worth of this accumulated biological matter every year, three to four times more than in 1956. While plants are a renewable resource, fossil carbon for our purposes is not. It will take millions of years to renew the earth's supply of coal and oil.

Before the industrial revolution, there were 580 billion tons of carbon in the atmosphere. Today there are 750 billion tons. That accumulation, the result of burning fossil fuels, is causing the climate change crisis. Humanity needs to solve this problem if we are to survive. It is the other carbon economy, the renewable carbon embodied in biodiversity, that offers the solution. Our dependence on fossil fuels has broken us out of nature's renewable carbon cycle. Our dependence on fossil fuels has fossilized our thinking.

Biodiversity is the alternative to fossil carbon. Everything that we derive from the petrochemical industry has an alternative in the realm of biodiversity. The synthetic fertilizers and pesticides, the chemical dyes, the sources of mobility and energy, all of these have sustainable alternatives in the plant and animal world. In place of nitrogen fertilizers, we have nitrogen-fixing leguminous crops and biomass recycled by earthworms (vermi-compost) or microorganisms (compost). In place of synthetic dyes, we have vegetable dyes. In place of the automobile, we have the camel, the horse, the bullock, the donkey, the elephant, and the bicycle.

Climate change is a consequence of the transition from biodiversity based on renewable carbon economies to a fossil fuel–based, nonrenewable carbon economy. This was the transition called the industrial revolution. While climate change, combined with peak oil and the end of cheap oil, is creating an ecological imperative for a post-oil, post–fossil fuel, postindustrial economy, the industrial paradigm is still the guiding force for the search for a transition pathway beyond oil. That's because industrialization has also become a cultural paradigm for measuring human progress. We want a post-oil world but do not have the courage to envisage a postindustrial world. As a result, we cling to the infrastructure of the energy-intensive fossil fuel economy and try to run it on substitutes such as nuclear power and biofuels. Dirty nuclear power is being redefined as "clean energy." Nonsustainable production of biodiesel and biofuel is being welcomed as a "green" option.

Humanity is playing these tricks with itself and the planet because we are locked into the industrial paradigm. Our ideas of the good life are based on production and consumption patterns that the use of fossil fuels gave rise to. We cling to these patterns without reflecting on the fact that they have become a human addiction only over the past fifty years and that maintaining this short-term, nonsustainable pattern of living for another fifty years comes at the risk of wiping out millions of species and destroying the very conditions for human survival on the planet. We think of well-being only in terms of human beings and, more accurately, only in terms of human beings over the next fifty years. We are sacrificing the rights of other species and the welfare of future generations.

To move beyond oil, we must move beyond our addiction to a certain model of human progress and human well-being. To move beyond oil, we must reestablish partnerships with other species. To move beyond oil, we must reestablish the other carbon economy, a renewable economy based on biodiversity.

Renewable carbon and biodiversity redefine progress. They redefine development. They redefine "developed," "developing," and "underdeveloped." In the fossil fuel paradigm, to be developed is to be industrialized—to have industrialized food and clothing, shelter and mobility, ignoring the social costs of displacing people from work and the ecological costs of polluting the atmosphere and destabilizing the climate. In the fossil fuel paradigm, to be underdeveloped is to have nonindustrial, fossil-free systems of producing our food and clothing, of providing our shelter and mobility. In the biodiversity paradigm, to be developed is to be able to leave ecological space for other species, for all people and future generations of humans. To be underdeveloped is to usurp the ecological space of other species and communities, to pollute the atmosphere, and to threaten the planet.

We need to change our mind before we can change our world. This cultural transition is at the heart of making an energy transition to an age beyond oil. What blocks the transition is a cultural paradigm that perceives industrialization as progress combined with false ideas of productivity and efficiency. We have been made to believe that industrialization of agriculture is necessary to produce more food. This is not at all true. Biodiverse ecological farming produces more and better food than the most energy- and chemical-intensive agriculture. We have been made to falsely believe that cities designed for automobiles provide more effective mobility to meet our daily needs than cities designed for pedestrians and cyclists.

Vested interests that gain from the sale of fertilizers and diesel, cars and trucks, have brainwashed us to believe that chemical fertilizers and cars mean progress. We have been reduced to buyers of their nonsustainable products rather than creators of sustainable, cooperative partnerships—both within human society and with other species and the earth as a whole.

The biodiversity economy is the sustainable alternative to the fossil fuel economy. The shift from fossil fuel–driven to biodiversity-supported systems reduces greenhouse gas emissions by emitting less and absorbing more CO_2. Above all, because the impacts of atmospheric pollution will continue even if we do reduce emissions, we need to create biodiverse ecosystems and economies because only they offer the potential to adapt to an unpredictable climate. And only biodiverse systems provide alternatives that everyone can afford. We need to return to the renewable carbon cycle of biodiversity. We need to create a carbon democracy so that all beings have their just share of useful carbon, and no one is burdened with carrying an unjust share of climate impacts due to carbon pollution.

Notes

1. Bloomberg, quoted by the BBC (April 14, 2008), in "Against the Grain: Making a Killing from Hunger," Grain.org, April 28, 2008.

2. BBC, "Action to Meet Asian Rice Crisis" (April 17, 2008), in "Against the Grain."

3. "Against the Grain."

4. Organization for Competitive Markets, "During a World Food Crisis, Monsanto Just Raised the Price of Corn Seed $100 a Bag," July 22, 2008.

5. Ernst Ulrich von Weizsäcker, Amory Lovins, and Hunter Lovins, *Factor Four: Doubling Wealth, Halving Resource Use* (London: Earthscan, 1997), 50.

6. Nicholas Stern, "Executive Summary (Full)," in *Stern Review: The Economics of Climate Change* (London, January 31, 2006), iv.

7. G. Barney, *The Global 2000: Report to the President* (Harmondsworth, UK: Penguin, 1980), cited in von Weizsäcker, Lovins, and Lovins, *Factor Four*.

8. "Energy Use in Organic Farming Systems," MAFF Project OFO 182, 1996–2000; R. Lai, "Soil Carbon Sequestration Impacts on Global Climate Change and Food Security," *Science* 304 (2004): 1623–27.

9. Paul Hepperly, *Organic Farming Sequesters Atmospheric Carbon and Nutrients in Soils* (Rodale Institute, October 15, 2003).

10. Caroline Lucas, Andy Jones, and Colin Hines, "Peak Oil and Food Security: Fuelling a Food Crisis," *Pacific Ecologist* 14 (Winter 2007): 13.

11. Dale Allen Pfeiffer, *Eating Fossil Fuels: Oil, Food, and the Coming Crisis in Agriculture* (Gabriola Island, BC: New Society, 2006), 20.

12. Ibid., 21.

13. *The Energy and Agriculture Nexus,* Environment and Natural Resources Working Paper no. 4 (Rome: FAO, 2000), 17.

14. Tim Appenzeller, "The End of Cheap Oil," *National Geographic,* June 2004.

15. Dannielle Murray, "Rising Oil Prices Will Impact Food Supplies," in *Fuelling a Food Crisis: The Impact of Peak Oil on Food Security,* ed. Caroline Lucas et al. (December 2006), 4.

16. Vandana Shiva, *The Violence of the Green Revolution* (London: Zed Books, 1991).

17. J. DeVries and G. Toenniessen, *Securing the Harvest: Biotechnology, Breeding and Seed Systems for African Crops* (Oxfordshire, UK: CABI, 2001), 36.

18. "Against the Grain."

19. Lucas et al., *Fuelling a Food Crisis.*

20. Louise E. Howard, *Sir Albert Howard in India* (London: Faber and Faber, 1953), xv.

21. Vandana Shiva, *Tomorrow's Biodiversity* (New York: Thames and Hudson, 2000).

22. Shiva, *Violence of the Green Revolution,* 114–16.

23. Pfeiffer, *Eating Fossil Fuels,* 24.

24. Stephen Bentley and Revenna Barker, "Fighting Global Warming at the Farmers Market," in *Food Share Research Action Report* (Toronto: Foodshare, 2005).

25. Mario Giampietro and David Pimentel, *The Tightening Conflict: Population, Energy Use and the Ecology of Agriculture* (NPG Forum Series, 1995).

26. Andy Jones, *Eating Oil: Food in a Changing Climate* (London: Sustain/ELM Farm Research Center, 2001).

27. Saritha Rai, "India to Import Wheat after Six-Year Hiatus," *New York Times,* June 30, 2006.

28. (Srinivas 2003).

29. Global Trade Information Services, Inc.

30. Jeremy Lawrence, "The Real Cost of a Bag of Salad: You Pay 99p. Africa Pays 50 Litres of Fresh Water," *Independent,* April 29, 2006.

31. Vandana Shiva et al., *Monocultures of the Mind: Perspectives on Biodiversity and Biotechnology* (London: Zed Books, 1993), 73.

32. David Pimentel, "Study Backs Operating Efficiency of Organic Sector," *Food Production Daily,* July 15, 2005.

33. Gemenez E. Holt, "Measuring Farmers' Agroecological Resistance after Hurricane Mitch in Nicaragua: A Case Study in Participatory, Sustainable Land Management Impact Monitoring," *Agriculture Ecosystems and Environment* 93 (2002): 87–105.

34. Laura Sayre, *Organic Farming Combats Global Warming—Big Time* (Rodale Institute, October 10, 2003); Nadia Scialabba and C. Hattam, *Organic Agriculture, Environment and Food Security* (Rome: FAO, 2002).

35. Vandana Shiva, *Biodiversity Based Organic Farming: A New Paradigm for Food Security and Food Safety* (New Delhi: Navdanya, 2006).

36. Edward Goldsmith, "How to Feed People under a Regime of Climate Change," *World Affairs Journal* (Winter 2003).

37. Charles Darwin, *The Formation of Vegetable Mould through the Action of Worms, with Observations on Their Habits* (London: Faber and Faber, 1927).

38. *Principles of Organic Farming* (New Delhi: Navdanya, 2006), 99.

39. Vandana Shiva, *Staying Alive: Women, Ecology, and Development* (London: Zed Books, 1988), 107–8.

40. Navdanya study on monocrop versus biodiverse production.

41. Shiva, *Biodiversity Based Organic Farming,* 27.

42. Debal Deb, *Industrial versus Ecological Agriculture* (New Delhi: Navdanya, 2004).

43. Akshat, *Rice Varieties of India* (New Delhi: Navdanya, 2006).

44. Kanak, *Wheats of India* (New Delhi: Navdanya, 2006) and *Forgotten Foods* (New Delhi: Navdanya, 2006).

45. ETC Group, "Gene Giants Grab 'Climate Genes': Annual Global Food Crisis, Biotech Companies Are Exposed as Climate Change Profiteers," ETC Group news release, May 13, 2008.

46. Greenpeace, *Eating Up the Amazon* (Netherlands: Greenpeace, April 2006).

47. Ibid., 5.

48. Felicity Lawrence, "Should We Worry about Soya in Our Food?" *Guardian*, July 25, 2006.

49. Michael Pollan, *The Omnivore's Dilemma: A Natural History of Four Meals* (New York: Penguin, 2006), 136–37.

50. Richard Heinberg, "The Essential Re-localization of Food Production," in *One Planet Agriculture: The Case for Action* (Bristol: Soil Association, UK, 2007), 13.

51. Rob Hopkins, "A Vision for Food and Farming in 2030," in Heinberg, *One Planet Agriculture*, 21–22.

13

The GMO Emperor
Has No Clothes

Genetic Engineering Is a Failed Technology

Technologies Are Tools

Technologies are tools. They are ways of doing, or making things. They are means of transforming what nature has given into food, clothing, shelter, means of mobility, means of communication. The word *technology* is derived from two Greek words: *techne,* which means tools, methods, means, and *logos,* which refers to thought or expression. As a tool, a technology is as good as the human ends it serves. It is not an end in itself. Yet technology has been elevated to a human end in our times.

Tools, methods, and means are assessed and evaluated on the basis of the ends they are meant to serve. They are assessed for their impact on nature, on society. Tools are compared with other alternative means available to do the job. And on the basis of an assessment, a choice is made to ensure there is no ecological or social harm, and that there is a positive contribution to human well-being. Until twenty years ago, before globalization and corporate rule, national governments and the UN had offices of technology assessment. The deregulation of corporate activity went hand in hand with the dismantling of technology assessment. When we stop perceiving technology as a means, mediating between nature and human needs, and falsely elevate it to an end, we falsely give it the status of a religion. The seeds of the Green Revolution, bred for responding to chemical fertilizers, were called "miracle seeds," and Norman Borlaug called the twelve people he sent across the world to spread chemicals through the introduction of the new seeds his "wheat apostles." This is the discourse of religion, not of science and technology.

When the Green Revolution was introduced into India in 1965–1966, no assessment was made about the impact of chemical fertilizer on soil organisms, soil structure, the water-holding capacity of the soils. No assessment was made to compare the yields of Green Revolution varieties and outputs of indigenous varieties and mixed-farming systems. When we started to conserve native seeds through the Navdanya movement in 1987, we found many of the indigenous varieties outperformed the Green Revolution varieties in grain yield, and most outperformed them in total biomass yield, which is what really counts since the grain is eaten by humans, but straw is food for the soil organisms and for farm animals. And our work on mixtures and biodiverse systems of farming shows that as a system, indigenous biodiversity produces more food and nutrition per acre. If we had a scientific approach to making choices about the technologies we use to produce our food, agroecology would win hands down. But the Green Revolution is promoted blindly as a religion, not on the basis of science.

Genetic engineering is the latest tool being imposed on India and the world as the new miracle. There are only three groups of GMO applications—Bt crops that are supposed to control pests, herbicide-resistant crops that are supposed to control weeds, and future promises of biofortification in the form of golden rice for addressing vitamin A deficiency and GMO bananas for removing iron deficiency.

When we assess genetic engineering as a tool aiming to achieve the objectives of reducing pests and weeds or increasing the availability of vitamin A and iron, it is clearly failing the test. GMOs have created superpests and superweeds instead of reducing pests and weeds. Golden rice is 7,000 percent less efficient in providing vitamin A and GMO bananas will be 3,000 percent less efficient in providing iron than alternatives available in our rich but rapidly disappearing biodiversity. GMOs continue to be promoted as a religion in spite of all the evidence that they are failing to do the job they are claimed to be designed for.

As in all religious fundamentalisms, there is intolerance of alternatives—alternative paradigms, alternative approaches to food production, and independent science. And we are witnessing the viciousness with which the industry attacks anyone who provides an alternative.

The new seed legislation introduced by the European Commission on May 6, 2013, is a desperate attempt by the biotechnology industry to criminalize the alternative of open-source seeds for farms and gardens in order to establish a monopoly of the seed and biotechnology industry. Another example is the attack on scientists whose scientific research has provided evidence of harm. The more those in the industry claim that the GMO debate is about science, the more they silence science and replace it with their pseudo-religion. Technological determinism replaces technological pluralism. Technological totalitarianism replaces democratic choice and responsibility.

Another consequence of making technology an end rather than a means is ignoring its impacts and failing to take responsibility for the harm to nature and people. The ultimate expression of irresponsibility is to create immunity for those who cause harm, recent examples being the Monsanto protection acts in the United States and in India (in the form of BRAI).

There is no science in DNA as a "master molecule" and genetic engineering as a game of Lego in which genes are moved around without any impact on the organism or the environment. This is a new pseudoscience that has taken on the status of a religion. There is no science justifying patents on life and seed. Shuffling genes is not making life. Living organisms make themselves. Patents on seed necessarily mean denying the contributions of millions of years of evolution and thousands of years of farmers' breeding. One could say that a new religion, a new cosmology, new creation myth is being put in place in which biotechnology corporations like Monsanto replace creation as "creators." GMO means "God, move over." Stewart Brand has actually said, "We are as gods and we had better get used to it."[1]

It is time to put nature and people back in the technology narrative. It is time to see technology as a tool, and not an end defining a new fundamentalist religion through which corporations become the new gods and the new "creators."

GMOs: A Failed Technology

We have been repeatedly told that genetically engineered crops (GE) will save the world. They will save the world by increasing yields and producing more food. They will save the world by controlling pests and weeds. They will save the world by reducing chemical use in agriculture. They will save the world by GE drought-tolerant seeds and other seed traits that will provide resilience in times of climate change. However, the GE emperor (Monsanto) has no clothes. All of these claims have been established to be false from years of experience all across the world. The Global Citizens Report *The Emperor Has No Clothes* brings together evidence from the ground of Monsanto's false promises and failed technology.

GMOs Fail to Address Hunger

The 2013 World Food Prize, which is sponsored by gene giants like Monsanto, was given to Monsanto by Monsanto. As Frances Moore Lappe and I wrote in a statement signed by awardees of the Alternative Nobel Prize, the Right Livelihood Award, and Councillors of the World Future Council, "This choice betrays the award's mandate to honor those contributing to a 'nutritious and sustainable food supply for all people.' GMO seeds are not designed for this purpose and function in ways that actually impede progress toward the goal."[2]

Contrary to its claim of feeding the world, genetic engineering has not increased the yield of a single crop. Navdanya's research in India has shown that notwithstanding Monsanto's assertion that Bt cotton yields fifteen hundred kilograms per acre, the reality is that the yield is an average of four hundred to five hundred kilograms per acre. Although Monsanto's Indian advertising campaign reports a 50 percent increase in yields for its Bollgard cotton, a survey conducted by the Research Foundation for Science, Technology, and Ecology found that the yields in all trial plots were lower than what the company promised. According to the Central Institute for Cotton Research (CICR), Nagpur, productivity of Bt cotton has been stagnant for the past five years. Bt cotton's failure to deliver higher yields has been reported all over the world. The Mississippi Seed Arbitration Council ruled that in 1997 Monsanto's Roundup Ready cotton failed to perform as advertised, recommending payments of nearly $2 million to three cotton farmers who suffered severe crop losses.

The report of the U.S.-based Union of Concerned Scientists entitled *Failure to Yield* has established that genetic engineering has not contributed to yield increases in any crop. According to this report, increases in crop yields in the United States are due to yield characteristics of conventional crops, not genetic engineering. As the University of Canterbury research team, led by Professor Jack Heinemann, has shown, North American crop production has fallen behind that of western Europe, despite U.S. farmers using genetically modified seed and more pesticide. The main point of difference between the regions is the adoption of GM seed in North America and the use of non-GM seed in Europe, the researchers say.[3]

As Marc Lappe and Britt Bailey report in their book *Against the Grain,* herbicide-resistant soybeans yielded 36 to 38 bushels per acre, while hand-tilled soybeans yielded 38.2 bushels per acre. According to the authors, this raises the possibility that the gene inserted into these engineered plants may selectively disadvantage their growth when herbicides are not applied. "If true, data such as these cast doubt on Monsanto's principal point that their genetic engineering is both botanically and environmentally neutral," the authors write.[4]

GMOs Fail to Address Malnutrition

Golden rice to remove vitamin A deficiency and end blindness and iron-enriched GMO bananas to prevent Indian women from dying in childbirth because of iron deficiency anemia are two of the nutritional promises from genetic engineering. Here, too, GMOs fail compared to alternatives. Nature has given us a cornucopia of biodiversity, rich in nutrients. Malnutrition and nutrient deficiency result from destroying biodiversity, and with it rich sources of nutrition. The Green Revolution has spread monocultures of chemical rice and wheat, driving out biodiversity from our farms and diets. And what survived as spontaneous crops like the amaranth greens and chenopodium (bathua), which are rich in iron, were sprayed with poisons and herbicides. Instead of being seen as iron-rich and vitamin-rich gifts, they were treated as "weeds." A Monsanto representative once said that the company's propriety herbicide Roundup killed the weeds that "steal the sunshine." And their Roundup ads in India tell women, "Liberate yourself; use Roundup." This is a recipe not for liberation but for being trapped in malnutrition.

As the "monoculture of the mind" took over, biodiversity disappeared from our farms and our food. The destruction of biodiverse-rich cultivation and diets has given us the malnutrition crisis, with 75 percent of women now suffering from iron deficiency.

Our indigenous biodiversity offers rich sources of iron. Amaranth has 11 milligrams per 100 grams of food, buckwheat has 15.5, neem 25.3, bajra 8, rice bran 35, rice flakes 20, Bengal gram roasted 9.5, Bengal gram leaves 23.8, cowpea 8.6, horse gram 6.77, amaranth greens up to 38.5, karonda 39.1, lotus stem 60.6, coconut meal 69.4, niger seeds 56.7, cloves 11.7, cumin seeds 11.7, mace 12.3, mango powder (amchur) 45.2, pippali 62.1, poppy seeds 15.9, tamarind pulp 17, turmeric 67.8, raisins 7.7. Bananas only have 0.44 milligrams of iron per 100 grams of edible portion. All the effort to increase the iron content of bananas will fall short of the iron content of our indigenous biodiversity. GMO bananas will be 3,000 percent less efficient than biodiversity alternatives in reducing iron deficiency anemia in Indian women.

The knowledge of growing this diversity and transforming it to food is women's knowledge. That is why in Navdanya we have created the network for food sovereignty in women's hands—Mahila Anna Swaraj.

The solution to malnutrition lies in growing nutrition, and growing nutrition means growing biodiversity. It means recognizing the knowledge of biodiversity and nutrition among millions of Indian women who have received it over generations as "grandmother's knowledge." For removing iron deficiency, iron-rich plants should be grown everywhere, on farms, in kitchen gardens, in community gardens, in school gardens, on rooftops, on balconies. Iron deficiency was not created by nature, and we can get rid of it by becoming cocreators and coproducers with nature.

But there is a "creation myth" that is blind to nature's creativity and biodiversity, and to the creativity, intelligence, and knowledge of women. According to this "creation myth" of capitalist patriarchy, rich and powerful men are the "creators." They can own life through patents and intellectual property. They can tinker with nature's complex evolution over millennia, and claim their trivial yet destructive acts of gene manipulation "create" life, "create" food, "create" nutrition. In the case of GM bananas, it is *one* rich man, Bill Gates, financing *one* Australian scientist, Dale, who knows *one* crop, the banana, imposing inefficient and hazardous GM bananas on millions of people in India and Uganda who have grown hundreds of banana varieties over thousands of years in additional to thousands of other crops.

The project is a waste of money and a waste of time. It will take ten years and millions of dollars to complete the research. But in the meantime, governments, research agencies, and scientists will become blind to biodiversity-based, low-cost, safe, time-tested, democratic alternatives in the hands of women.

GMO Crops Fail to Control Pests and Weeds or Reduce the Use of Toxic Chemicals

In twenty years of commercialization of GE crops, only two traits have been commercialized on a significant scale: herbicide tolerance and insect resistance (Bt crops). Herbicide tolerant, or Roundup Ready, crops were supposed to control weeds, and Bt crops were intended to control pests. Instead of controlling weeds and pests, GE crops have led to the emergence of superweeds and superpests. In the United States, Roundup Ready crops have produced weeds resistant to Roundup. Approximately 15 million acres are now overtaken by superweeds, and in an attempt to kill these weeds, farmers have been paid $12 per acre by Monsanto to spray more lethal herbicides such as Agent Orange, which was used during the Vietnam War.

In India, Bt cotton, sold under the trade name Bollgard, was supposed to control the bollworm pest. Today, the bollworm has become resistant to Bt cotton, and now Monsanto is selling Bollgard II, with two additional toxic genes in it. New pests have emerged, and farmers are using more pesticides. Studies carried out by Navdanya and the Research Foundation for Science, Technology and Ecology have shown that pesticide use in Vidharba in Maharashtra increased thirteenfold after the introduction of Bt cotton.

A study by Charles Benbrook reports that herbicide-resistant crop technology has led to a 239 million kilogram (527 million pound) increase in herbicide use in the United States between 1996 and 2011, while Bt crops have reduced insecticide applications by 56 million kilograms (123 million pounds). Overall, pesticide use increased by an estimated 183 million kilograms (404 million pounds), or about 7 percent.[5]

The study concludes, "Contrary to often-repeated claims that today's genetically-engineered crops have, and are reducing pesticide use, the spread of glyphosate-resistant weeds in herbicide-resistant weed management systems has brought about substantial increases in the number and volume of herbicides applied. If new genetically engineered forms of corn and soybeans tolerant of 2,4-D are approved, the volume of 2,4-D sprayed could drive herbicide usage upward by another approximate 50%. The magnitude of increases in herbicide use on herbicide-resistant hectares has dwarfed the reduction in insecticide use on Bt crops over the past 16 years, and will continue to do so for the foreseeable future."

Despite claims that GMOs will lower the levels of chemicals (pesticides and herbicides) used, this has not been the case. This is of great concern both because of the negative impacts of these chemicals on ecosystems and humans and because there is the danger that increased chemical use will cause pests and weeds to develop resistance, requiring even more chemicals in order to manage them. In India:

- A survey conducted by Navdanya in Vidharba showed that pesticide use has increased thirteenfold there since Bt cotton was introduced.
- A study recently published in the *Review of Agrarian Studies* also showed a higher expenditure for small farmers on chemical pesticides for Bt cotton than for other varieties.[6]
- Nontarget pest populations in Bt cotton fields have exploded, which will likely counteract any decrease in pesticide use.[7]

In China, where Bt cotton is widely planted:

- Populations of mirid bugs, pests that previously posed only a minor problem, have increased twelvefold since 1997. A 2008 study in the *International Journal of Biotechnology* found that any financial benefits of planting Bt cotton had been eroded by the increasing use of pesticides needed to combat nontarget pests.[8]

In the United States, due mainly to the widespread use of Roundup Ready seeds:

- Herbicide use increased 15 percent (318 million additional pounds) from 1994 to 2005—an average increase of a quarter pound per each acre planted with GM seed—according to a 2009 report published by the Organic Center.[9]
- The same report found that in 2008, GM crops required 26 percent more pounds of pesticides per acre than acres planted with conventional varieties, and projects that this trend will continue due to the spread of glysophate-resistant weeds.[10]
- Moreover, the rise of glysophate (the herbicide in Roundup) resistant weeds has made it necessary to combat these weeds by employing other, often more toxic, herbicides. This trend is confirmed by 2010 USDA pesticide data, which show skyrocketing glysophate use accompanied by constant or increasing rates of use for other, more toxic, herbicides.[11]
- Moreover, the introduction of Bt corn in the United States has had no impact on insecticide use, and while Bt cotton is associated with a decrease in insecticide use in some areas, insecticide applications in Alabama, where Bt cotton is planted widely, doubled between 1997 and 2000.[12]

In Argentina, after the introduction of Roundup Ready soy in 1999:

- Overall glysophate use more than tripled by 2005–2006. A 2001 report found that Roundup Ready soy growers in Argentina used more than twice as much herbicide as conventional soy growers.[13]
- In 2007, a glysophate-resistant version of Johnsongrass (considered one of the worst and most difficult weeds in the world) was reported on over 120,000 hectares of prime agricultural land—a consequence of the increase in glysophate use. As a result, it was recommended that farmers use a mix of herbicides other than glysophate (often more toxic) to combat the resistant weeds, and it is estimated that an additional twenty-five liters of herbicides will be needed each year to control the resistant weeds.[14]

In Brazil, which has been the world's largest consumer of pesticides since 2008:[15]

- GM crops became legally available in 2005, and now make up 45 percent of all row crops planted in Brazil—a percentage that is expected to increase.[16]
- Soy area has increased 71 percent, but herbicide use has increased 95 percent.[17]
- Of eighteen herbicide resistant weed species reported, five are glysophate resistant.[18]
- In 2009, total herbicide active ingredient use was 18.7 percent higher for GM crops than for conventional ones.[19]

Patents on Seeds and Seed Monopolies

GMOs are intimately linked to seed patents. In fact, patenting of seeds is the real reason why industry is promoting GMOs. Monopolies over seeds are being established through patents, mergers, and cross-licensing arrangement. Monsanto now controls the world's biggest seed company, Seminis, which has bought up Peto Seed, Bruinismo, Genecorp, Barhan, Horticere, Agroceres, Royal Suis, Choon Ang, and Hungnong. Other seed acquisitions and joint ventures of Monsanto are Asgrow, De Rinter, Monsoy, FT Sementes, Carma, Advanta Canola, China Seed, CNDK, ISG, Wertern, Protec, Calgene, Deltapine Land, Syngenta Global Cotton Division, Agracetus, Marneot, EID Parry Rallis, CDM Mandiyu, Ciagro, Renessan, Cargill, Terrazawa, Cargill International Seed Division, Hybritech, Jacob Hartz 1995, Agriprowheat, Cotton States, Limagrain Canada, Alypanticipacoes, First Line, Mahyco, Corn States Intl, Corn States Hybrid, Agroeste, Seusako, Emergent Genetics, Mahendra, Indusem, Darhnfeldt, Paras, Unilever, Dekelb, Lustum, Farm Seed, Deklbayala, Ayala, Polon, Ecogen, and PBIC. In addition, Monsanto has cross-licensing arrangements with BASF, Bayer, Dupont, Sygenta, and Dow. They have agreements to share patented genetically engineered seed traits with each other. The giant seed corporations are not competing with each other. They are competing with peasants and farmers over the control of the seed supply.

The combination of patents, genetic contamination, and spread of monocultures means that society is rapidly losing its seed freedom and food freedom. Farmers are losing their freedom to have seed and grow organic food free of the threat of contamination by GE crops. Citizens are losing their freedom to know what they are eating, and to have the choice to eat GE-free food.

An example of seed monopolies is cotton in India. After Monsanto gained control of 95 percent of the cottonseed market, seed prices jumped 8,000 percent. India's antitrust court, the Monopoly and Restrictive Trade Practices Commission, was forced to rule against Monsanto. High costs of seed and chemicals have pushed 250,000 farmers to suicide, with most suicides concentrated in the cotton belt.

Monsanto does not control the seed only through patents. It also spreads its control through contamination. After spreading genetic contamination, Monsanto sues farmers as "intellectual property thieves," as it did in the case of Percy Schmeiser. That is why a case has been brought against Monsanto by a coalition of more than eighty groups to stop the company from suing farmers after polluting their crops.[20]

GMOs and Seeds of Suicide

The announcement on Monsanto India's website declares, "Monsanto is an agricultural company. We apply innovation and technology to help farmers around the world produce more while conserving more. Producing more, conserving more, improving farmers' lives." All the pictures are of smiling prosperous farmers from the state of Maharashtra. However, the reality on the ground is completely different. Farmers who have become dependent on Monsanto's seed monopoly are in debt and in deep distress. Most of the farmers who have committed suicide in India due to being trapped in debt are in the cotton belt, which has become a suicide belt. The highest number of suicides is in Maharashtra. And 95 percent of the cottonseed is now controlled by Monsanto. Monsanto's talk of "technology" tries to hide its real objectives of ownership and control over seed; genetic engineering is just a means to control seed and the food system through patents and intellectual property rights.

A Monsanto representative admitted that the company was "the patient, diagnostician, and physician all in one" in writing the patents on life sections in the TRIPS agreement of WTO. Stopping farmers from saving seeds and exercising their seed sovereignty was the objective. And Monsanto has gone very far down the road of destroying biodiversity and farmers' seed sovereignty. It is now extending its patents to conventionally bred seed, as in the case of broccoli and capsicum, or the low-gluten wheat it had pirated from India, which we challenged as a biopiracy case in the European Patent Office.[21]

An epidemic of farmers' suicides has spread across four states of India over the last decade. According to official data, more than 284,694 farmers have committed suicide in India since 1995. These four states are Maharashtra, Andhra Pradesh, Karnataka, and Punjab. The suicides are most frequent where farmers grow cotton and have been a direct result of the creation of seed monopolies.

Increasingly, the supply of cottonseeds has slipped out of the hands of farmers and the public system into the hands of global seed corporations like Monsanto. The entry of seed MNCs was part of the globalization process. Corporate seed supply implies a number of shifts simultaneously.

First, giant corporations start to control local seed companies through buyouts, joint ventures, and licensing arrangements, leading to a seed monopoly. The entry of Monsanto in the Indian seed sector was made possible with a 1988 seed policy, imposed by the World Bank, requiring the government of India to deregulate the seed sector. Indian companies were locked into joint ventures and licensing arrangements, and concentration over the seed sector increased. In the case of cotton, Monsanto now controls 95 percent of the cottonseed market through its GMOs.

Second, seed is transformed from being a common good to being the "intellectual property" of Monsanto, for which the corporation can claim limitless profits through royalty payments. For the farmer this means deeper debt.

Third, seed is transformed from a renewable regenerative, multiplicative resource into a nonrenewable resource and commodity. Seed scarcity is a consequence of seed monopolies, based on the nonrenewability of seed: beginning with hybrids, moving to genetically engineered seed like Bt cotton, progressing to the ultimate aim of the "terminator" seed, which is engineered for sterility. Each of these technologies of nonrenewability is guided by one factor alone—forcing farmers to buy seed every planning season. For farmers this means higher costs. For seed corporations it translates into higher profits.

Fourth, cotton, which had earlier been grown as a mixture with food crops, now had to be grown as a monoculture, with higher vulnerability to pests, disease, drought, and crop failure.

Fifth, Monsanto started to subvert India's regulatory processes, and in fact started to use public resources to push its nonrenewable hybrids and GMOs through so-called public private partnerships (PPP). The field data of Bt cotton are also manipulated; when cotton yields are shown to be higher than in the pre-Bt cotton years, it is not mentioned that cotton has traditionally not been grown as a monoculture but as a mixed crop. Converting biodiversity to monocultures of course leads to an increase in the "yield" of the monoculture, but this is accompanied by a decline in production at the biodiversity level.

Sixth, the creation of seed monopolies is based on the simultaneous deregulation of seed corporations, including biosafety and seed deregulation, and super-regulation of farmers' seeds and varieties. Globalization allowed seed companies to sell self-certified seeds, and in the case of genetically engineered seed, they are seeking self-regulation for biosafety. This is the main aim of the recently proposed Biotechnology Regulatory Authority of India bill, which I have named the "Monsanto Protection Act" and which is in effect a biosafety deregulation authority. The proposed Seed Bill 2004, which has been blocked by a massive nationwide Gandhian seed satyagraha by farmers, aimed at forcing all farmers to register the varieties they have evolved over millennia. This compulsory registration and licensing system robs farmers of their fundamental freedoms. Such laws are being introduced in every country.

The creation of seed monopolies and with it the creation of unpayable debt to a new species of moneylender, the agents of the seed and chemical companies, has led to hundreds of thousands of Indian farmers killing themselves since 1997.

The creation of seed monopolies, the destruction of alternatives, the collection of superprofits in the form of royalties, and the increasing vulnerability of monocultures have bred a context for debt, suicides, and agrarian distress.

I have always been critical of reductionism. I look at systems and at contextual causation. It is this *system* that Monsanto has created of seed monopoly, crop monocultures, and a context of debt, dependency, and distress that is driving the farmers' suicide epidemic in India. This systemic control has been intensified with Bt cotton. That is why most suicides are in the cotton belt. The suicides first started in the district of Warangal in Andhra Pradesh. Peasants in Warangal used to grow millets, pulses, and oilseeds. Overnight, Warangal was converted to a cotton-growing district based on nonrenewable hybrids that need irrigation and are prone to pest attacks. Small peasants without capital were trapped in a vicious cycle of debt. Some ended up committing suicide.

This was the period when Monsanto and its Indian partner Mahyco were also carrying out illegal field experiments with genetically engineered Bt cotton. All imports and field trials of genetically engineered organisms in India are governed by a law under the Environment Protection Act called the Rules for the Manufacture, Use, Import, Export and Storage of Hazardous Microorganisms, Genetically Engineered Organisms or Cells 1989. We at the Research Foundation for Science, Technology, and Ecology used these laws to stop Monsanto's commercialization of Bt cotton in 1999, which is why approval was not granted for commercial sales until 2002. The government of Andhra Pradesh filed a case in the Monopoly and Restrictive Trade Practices Act (MRTP), India's antitrust law, arguing that Monsanto's seed monopolies were the primary cause of farmers' suicides in Andhra Pradesh. Monsanto was forced to reduce its prices of Bt cottonseeds. The high costs of seeds and other inputs were combined with falling prices of cotton due to a $4 billion U.S. subsidy and the dumping of this subsidized cotton on India by using the WTO to force India to remove quantitative restrictions on agricultural imports. Rising costs of production and falling prices of the product is a recipe for indebtedness and is the main cause of farmers' suicides. This is why farmers' suicides are most prevalent in the cotton belt, which by seed industries' own claim is rapidly becoming a Bt cotton belt. Bt cotton is thus heavily implicated in farmers' suicides.

The technology of engineering Bt genes into cotton was aimed primarily at controlling pests. However, new pests have emerged in Bt cotton, leading to higher use of pesticides. In the Vidharba region of Maharashtra, which has the highest suicides, the area under Bt cotton has increased from 0.200 million hectares in 2004 to 2.880 million hectares in 2007. Costs of pesticides for farmers have increased from Rs. 921 million to Rs. 13,264 billion in the same period, which is a thirteenfold increase. A pest-control technology that fails to control pests might be good for seed corporations, which are also agrochemical corporations. For farmers it translates into suicide.

Monsanto and its PR men are trying desperately to delink the epidemic of farmers' suicides in India from its growing control over the cottonseed supply. For us it is the control over seed, the first link in the food chain, the source of life, which is our biggest concern. When a corporation controls seed, it controls life, including the lives of our farmers.

The trends of Monsanto's concentrated control of the seed sector in India and across the world is the central issue. This is what connects the farmers' suicides in India to *Monsanto vs. Percy Schmeiser* in Canada or *Monsanto vs. Bowman* in the United States to farmers in Brazil suing Monsanto for $2.2 billion for unfair collection of royalties. Through patents on seed Monsanto has become the "life lord" of the planet, collecting rents from life's renewal and from farmers, the original breeders. Patents on seed are illegitimate because putting a toxic gene into a plant cell is not the "creation" or invention of the plant. They are seeds of deception—the deception of Monsanto being the creator of seeds and life, the deception that while it sues farmers and traps them in debt it is working for farmers' welfare and "improving farmers' lives," the deception that GMOs feed the world. The deception about the emperor's new clothes.

In 1995, Monsanto introduced its Bt technology in India through a joint venture with the Indian company Mahyco. In 1997–1998, Monsanto illegally started open field trials of its propriety GMO Bt cotton and announced it would be selling the seeds commercially the following year. India has rules for regulating GMOs since 1989 under the Environment Protection Act. Under these rules it is mandatory to get approval from the Genetic Engineering Approval Committee under the Ministry of Environment for GMO trials. When we found out that Monsanto had not applied for approval, the Research Foundation for Science, Technology, and Ecology sued Monsanto in the Supreme Court of India. As a result Monsanto could not start commercial sales of its Bt cottonseeds until 2002. But it had started to change Indian agriculture before that.

Recently Monsanto has been publishing news articles that propagate lies and false claims about the yield and prosperity achieved by Bt cotton. One such article, "Farmers Reaped Gold through Bt Cotton," was published in the *Times of India* on October 31, 2008, and this again was repeated on August 28, 2011. The article says, "The switch over from conventional cotton to Bt cotton in the villages [Bhamraja and Antargaon] has led to social and economic transformation. There are no suicides and people are prospering in agriculture." But the visit by Navdanya to Bhamraja and Antargaon showed that no farmer had reaped gold through Bt cotton. Whatever little success some farmers had achieved, it was through some other sources. But Monsanto claimed these successes were achieved by Bt cotton. The news article claims that since the adoption of Bt cotton, there have been no farmers' suicides in Bhamraja—though villagers reported fourteen such suicides.

In another advertisement, Monsanto claimed that the company's "Bt cotton seeds had helped to create additional income of over Rs. 31,500 crore for 60 lakh cotton farmers by reducing pesticide use and increasing yield." The Advertising Standards Council of India (ASCI) found that the claims were baseless, unsubstantiated by facts and figures. The ASCI asked Monsanto to drop the claims and the advertisement, to which it agreed.

Faced with severe criticism of Bt cotton all over the world, Monsanto and other multinational seed companies are making desperate and futile attempts by funding articles, reports, and reviews that promote Bt cotton and conceal the grim scenario of farmers' suicides and indebtedness due to the failure of Bt cotton.

Earlier, too, the International Food Policy Research Institute (IFPRI), the International Service for the Acquisition of Agribiotech Applications (ISAAA), the Associated Chambers of Commerce and Industry (ASSOCHAM), and the Indian Market Research Bureau (IMRB) had published reports that contained not one iota of truth. An article titled "Case Studies: A Hard Look at GM Crop," by Natasha Gilbert, published in Nature on May 1, 2013, tried to deny links between farmers' suicides, GMOs, and seed monopolies. A report by the IFPRI states, "In specific regions and years, where Bt-cotton may have indirectly contributed to farmer indebtedness (via crop failure) leading to suicides, its failure was mainly the result of the context or environment in which it was introduced or planted; Bt-cotton as a technology is not to blame."

This is an interesting argument. A technology is always developed in the context of local socioeconomic and ecological conditions. A technology that is a misfit in a context is a failed technology for that context. You cannot blame the context for a failed technology.

Monsanto and other seed companies have been spreading false propaganda that Bt cotton is not responsible for the farmers' suicides in Vidarbha. To unravel the truth, Navdanya conducted a study in Vidarbha from February 13 to February 25, 2009, covering four districts—Yavatmal, Wardha, Amrawati, and Washim. The study found that 84 percent of farmers' suicides were attributed to Bt cotton failure.

India has witnessed more than 284,694 farmers' suicide in a span of seventeen years, from 1995 to 2012; however, worst is the case of Maharashtra, the state that today has the highest acreage under Bt cotton. In the state there were 1,083 farmer suicides in 1995, which increased to 3,695 in 2002, more than a three times' jump, coinciding with Monsanto's introduction of Bt cotton. The situation in Vidarbha is more grim. There were only fifty-two farmer suicides in 2001, but since 2002 the number has increased alarmingly. Here are the statistics of farmer suicides over the years in Vidarbha:

2001	52
2002	104
2003	148
2004	447
2005	445
2006	1,148
2007	1,246
2008	1,248
2009	916
2010	748
2011	916
2012	927

The figures hide lives ruined as collateral damage. Every suicide ruins the lives of eight to nine people in a family. A simple calculation shows that during 2002–2011 the lives of 55,000–65,000 people were ruined due to the farmers' suicides in Vidarbha. The stories of surviving family members are tragic. With the husband's death, a new vicious cycle of debt is set in motion as the widow inherits her husband's debts and works round the clock to pay back as well as make ends meet.

According to P. Sainath, who has covered farmers' suicides systematically, "The total number of farmers who have taken their own lives in Maharashtra since 1995 is closing in on 54,000. Of these 33,752 have occurred in nine years since 2003, at an annual average of 3,750. The figure for 1995–2002 was 20,066 at an average of 2,508." Suicides have *increased* since Bt cotton was introduced. The price of seed jumped 8,000 percent. Monsanto's royalty extraction and the high costs of purchased seed and chemicals have created a debt trap. According to government of India data, nearly 75 percent of rural debt is due to purchased inputs. Farmers' debt grows as Monsanto profits grow. It is in this systemic sense that Monsanto's seeds are seeds of suicide. An internal advisory by the Agricultural Ministry of India in January 2012 had this to say of the cotton-growing states in India—"Cotton farmers are in a deep crisis since shifting to Bt cotton. The spate of farmer suicides in 2011–12 has been particularly severe among Bt cotton farmers."[22]

Recent data for the year 2012, released by National Crime Records Bureau (NCRB), present a more worrying scenario for farmers' suicides in the country. Figures for eighteen years, 1995–2012, show that at least 284,694 farmers have committed suicide, excepting in Chhattisgarh and West Bengal, for which no figures are available.

The situation is worst in Maharashtra and Andhra Pradesh. Both are main cotton-producing states where more than 95 percent of acreage is covered by Bt cotton. In Maharashtra, farmers' suicides jumped sharply to 3,786 in 2012 from 3,337 in 2011, an increase of 449, the worst annual increase in the last seven years. Andhra Pradesh also witnessed an upward trend, from 2,206 in 2011 to 2,572 in 2012, 366 more than the previous year.

In the nine-year period 1995–2003, India recorded 138,321 farmers' suicides, an annual average of 15,369. For 2004–2012, the figure is 146,373, an annual average of 16,264. This means the second nine years saw a higher annual average than the first nine years. The following presents the total of farmers' suicides in all of India in the period 1995–2012:

1995	10,720
1996	13,729
1997	13,622
1998	16,015
1999	16,082
2000	16,603

2001	16,415
2002	17,971
2003	17,164
2004	18,241
2005	17,131
2006	17,060
2007	16,632
2008	16,196
2009	17,368
2010	15,964
2011	14,027
2012	13,754
Total	284,694

In light of the farmers' suicides, the Parliamentary Committee on Agriculture has called for a ban on GMO crops. The panel of technical experts appointed by the Supreme Court has recommended a ten-year moratorium on field trials of all GM food and termination of all ongoing trials of transgenic crops.[23] And the ultimate seeds of suicide come in the form of Monsanto's patented terminator technology to create sterile seed, or suicide seeds. The Convention on Biological Diversity has banned its use; otherwise Monsanto would be collecting even higher profits from seed.

There are alternatives to Bt cotton and toxic pesticides. Through Navdanya we have promoted organic farming and seeds of hope to help farmers move away from Monsanto's seeds of suicide. Organic farmers in Vidharba are earning Rs. 6,287 per acre on average, compared to Bt cotton farmers, who are earning Rs. 714 per acre on average. Many Bt cotton farmers have a negative income, hence the suicides.

That is why we have started Fibres of Freedom in the heart of Monsanto's Bt cotton suicide belt in Vidharba. We have created community seed banks with indigenous seeds and helped farmers go organic. No GMO seeds, no debt, no suicides. We save and share seeds of life and seeds of freedom—diverse, open-pollinated, GMO-free, patent-free seeds. And we have started the Global Citizens Seed Freedom campaign (Seed Freedom: www.Navdanya.org; www.seedfreedom.in).

Technologies are tools. When the tool fails, it needs replacing. Bt cotton technology has failed to control pests or secure farmers' lives and livelihoods. It is time to replace GM technology with ecological farming. It is time to stop farmers' suicides.

Notes

1. Stewart Brand, *Whole Earth Discipline: Why Dense Cities, Nuclear Power, Transgenic Crops, Restored Wildlands, and Geoengineering Are Necessary* (New York: Penguin, 2010).

2. http://www.huffingtonpost.com/frances-moore-lappe-and-anna-lappe/choice-of-monsanto-betray_b_3499045.html; http://opinionator.blogs.nytimes.com/2013/06/25/the-true-deservers-of-a-food-prize/?hp&_r=0.

3. http://www.foodprocessing.com.au/news/61647-Do-GM-crops-equal-better-yields-UC-researchers-say-no.

4. Britt Bailey and Marc Lappe, *Against the Grain: Biotechnology and the Corporate Takeover of Your Food* (Monroe, ME: Common Courage, 2002).

5. Charles Benback, "Impacts of Genetically Engineered Crops on Pesticide Use in the U.S.—The First Sixteen Years," *Environmental Sciences Europe* (2012), 24:24. doi:10.1186/2190-4715-24-24.

6. Madhura Swaminathan and Vikas Rawal, "Are There Benefits from the Cultivation of Bt Cotton?" *Review of Agrarian Studies* 1, no. 1 (2011).

7. Glenn Davis Stone, "Field versus Farm in Warangal: Bt Cotton, Higher Yields, and Larger Questions," *World Development* 39, no. 3 (2011): 387.

8. GM Watch, "Benefits of Bt Cotton Elude Farmers in China," http://www.gmwatch.org/latest-listing/1-news-items/13089.

9. http://www.organic-center.org/science.pest.php?action=view&report_id=159.

10. http://www.organic-center.org/science.pest.php?action=view&report_id=159.

11. GM Watch, "Despite Industry Claims, Herbicide Use Fails to Decline with GM Crops," http://www.gmwatch.org/latest-listing/1-news-items/13089.

12. Charles Benbrook, "Do GM Crops Means Less Pesticide Use?" *Pesticide Outlook,* October 2001, http://www.biotech-info.net/benbrook_outlook.pdf.

13. Friends of the Earth International, "Who Benefits from GM Crops? Feed the Biotech Giants, Not the World's Poor," February 2009, http://www.foei.org/en/resources/publications/pdfs/2009/gmcrops2009exec.pdf.

14. Ibid.

15. GM Watch, "Use of Pesticides in Brazil Continues to Grow," April 18, 2011, http://www.gmwatch.org/latest-listing/1-news-items/13072-use-of-pesticides-in-brazil-continues-to-grow.

16. Soybean and Corn Advisor, "Brazilian Farmers Are Rapidly Adopting Genetically Modified Crops," March 10, 2010, http://www.soybeansandcorn

.com/news/Mar10_10-Brazilian-Farmers-Are-Rapidly-Adopting-Gentically-Modified-Crops.

17. BioWatch South Africa, "GM Agriculture: Promises or Problems for Farming in South Africa?" May 16 2011, http://www.sacau.org/hosting/sacau/SacauWeb.nsf/SACAU%202011_Biowatch-%20GM%20agriculture%20Promises%200r%20problems%20for%20farming%20in%20South%20Africa.pdf.

18. GM Watch, "Use of Pesticides in Brazil Continues to Grow."

19. Graham Brookes and Peter Barfoot, *GM Crops: Global Socio-economic and Environmental Impacts, 1996–2009* (PG Economics, 2011).

20. http://www.pubpat.org/assets/files/seed/OSGATA-v-Monsanto-Complaint.pdf.

21. http://www.no-patents-on-seeds.org/en/information/background/green-light-for-patents-on-plants-and- animals.

22. http://www.hindustantimes.com/News-Feed/Business/Ministry-blames-Bt-cotton-for-farmer-suicides/Article1-830798.aspx.

23. *Daily Mail,* October 18, 2012.

Appendix

Tables

Some of the tables expressing data used in chapter 9 are too large to print legibly on the pages of this book. Tables 9.1, 9.8, 9.9, and 9.10 are available at the following websites:
www.kentuckypress.com
www.navdanya.org

Table 2.1. The Consultative Group on International Agricultural Research (CGIAR) system, 1984

Acronym (year established)	Center	Location	Budget (millions of dollars)
IRRI (1960)	International Rice Research Institute	Los Baños, Laguna, Philippines	22.5
CIMMYT (1966)	Centro internaccional de mejoramientio maizy trigo	Mexico City, Mexico	21
IITA (1967)	International Institute of Tropical Agriculture	Ibadan, Nigeria	21.2
CIAT (1968)	Centro internacional de agricultura tropical	Cali, Colombia	23.1
CIP (1971)	Centro internacional de la papa	Lima, Peru	10.9
WARDA (1971)	West African Rice Development Association	Monrovia, Liberia	2.9
ICRISAT (1972)	International Crops Research Institute for the Semi-arid Tropics	Hyderabad, India	22.1
ILRAD (1973)	International Laboratory for Research for Animal Diseases	Nairobi, Kenya	9.7
IBPGR (1974)	International Board for Plant Genetic Resources	Rome, Italy	3.7
ILCA (1974)	International Livestock Center for Africa	Addis Ababa, Ethiopia	12.7
IFPRI (1975)	International Food Policy Research Institute	Washington, DC, USA	4.2
ICARDA (1976)	International Center for Agricultural Research in the Dry Areas	Aleppo, Syria	20.4
ISNAR (1980)	International Service for National Agricultural Research	The Hague, Netherlands	3.5

Source: Consultative Group on International Agricultural Research, Washington, DC, 1984.

Table 2.2. Membership of the Consultative Group on International Agricultural Research, January 1983

Countries	International organizations	Foundations	Fixed-term members representing developing countries
Australia	African Development Bank	Ford Foundation	Asian region: Indonesia and Pakistan
Belgium	Arab Fund for Economic Development	Research and Development Research Centre	African region: Senegal and Tanzania
Brazil	Asian Development Bank	Kellogg Foundation	Latin American region: Colombia and Cuba
Canada	Commission of the European Communities	Leverhulme Trust	Southern and eastern European region: Greece and Romania
Denmark	Food and Agriculture organization of the United Nations	Rockefeller Foundation	Near Eastern region: Iraq and Libya
France	Inter-American Development Bank		
Germany	International Bank for Reconstruction and Development		
India	OPEC Fund		
Ireland	United Nations Development Programme		
Italy	United Nations Environment Programme		
Japan			
Mexico			
Netherlands			
Nigeria			
Norway			
Philippines			
Saudi Arabia			
Spain			
Sweden			
Switzerland			
United Kingdom			
United States			

Source: Consultative Group on International Agricultural Research, Washington, DC.

Table 2.3. IRRI finances according to source (1961–80) (U.S. dollars)

Contributor	Amount	% of total	Year(s) of grant
Ford Foundation	23,950,469	18.84	1961–80
Rockefeller Foundation	20,460,431	16.1	1961–80
US AID	28,982,114	22.8	1967–80
International organizations	20,334,788	16	
Asian Development Bank	800,000		1975, 1977
European Economic Community	3,011,219		1978–80
Fertilizer Development Center	70,939		1979–80
Foundation for International Potash Research	7,375		1963–65
International Board for Plant Genetic Resources	208,100		1977, 1979–80
International Center of Insect Physiology and Ecology	125,432		1978–80
International Development Association	7,775,000		1973–80
International Development Research Center	3,710,736		1972–73, 1975–76, 1978–80
International Fund for Agricultural Research	500,000		1980
International Potash Institute / Potash Institute of North America	68,064		1963, 1965–66, 1968–69, 1971–79
OPEC Special Fund	200,000		1980
UN Development Program	3,559,273		1974–78, 1978
UN Economic and Social Commission	6,000		1970, 1979
UN Environment Program	280,000		1974–78
UN Food and Agriculture Organization (FAO)	2,650		1969
World Phosphate Rock Institute	10,000		1975
National governments	31,920,619	25.11	
Australia	4,185,459		1975–80
Belgium	148,677		1977
Canada	6,507,862		1974–80
Denmark	443,048		1978–80
Federal Republic of Germany	3,459,159		1974–80
Indonesia	1,619,119		1973–80
Iran	250,000		1977

(cont.)

Table 2.3. IRRI finances according to source (1961–80) (U.S. dollars) (cont.)

Contributor	Amount	% of total	Year(s) of grant
Japan	8,882,145		1971–77, 1979–80
Korea	82,259		1980
Netherlands	1,168,673		1971–1979
New Zealand	137,450		1973, 1976–78
Philippines	100,000		1980
Saudi Arabia	274,300		1976–77, 1980
Sweden	285,700		1979–80
United Kingdom	4,073,824		1973–76, 1979–80
Corporations	345,726	0.27	
Bayer	9,333		1971, 1973
Boots	1,000		1977
Chevron Chemicals	2,993		1972, 1977
Ciba-Geigy	20,500		1968, 1970, 1972, 1975, 1978–80
Cyanamid	19,000		1975–76, 1978, 1980
Dow Chemical	10,153		1967–70
Eli Lilly & Co. (ELANCO)	6,000		1968–70
Esso Engineering and Research	4,406		1964–68
FMC	9,000		1975–77, 1980
Gulf Research and Development	3,500		1969, 1972
Hoechst	11,891		1972, 1975–78
Imperial Chemical Industries	55,000		1967, 1969, 1971–76, 1979–80
International Business Machines (IBM)	7,000		1967
International Minerals and Chemical	60,000		1966–67, 1975
Kemanober	500		1980
Minnesota Mining and Manufacturing	1,000		1974
Monsanto	12,500		1967, 1969, 1971–72, 1976, 1978–80
Montedison	8,982		1977–78, 1980
Occidental Chemical	500		1971
Pittsburgh Plate Glass	2,000		1967
Plant Protection	5,000		1966
Shell Chemical	42,672		1969–70, 1972–73, 1975, 1977–78, 1980

Table 2.3. IRRI finances according to source (1961–80) (U.S. dollars) (cont.)

Contributor	Amount	% of total	Year(s) of grant
Stauffer Chemical	40,000		1967–69, 1971–76, 1978–80
Union Carbide	11,000		1968, 1970
Uniroyal Chemical	496		1980
Upjohn	1,200		1972
Government agencies	1,030,872	0.81	
National Food and Agriculture Council (Philippines)	276,859		1973, 1976–80
National Institute of Health (U.S.)	383,708		1978–80
National Science Development Board (Philippines)	104,172		1963–68, 1973, 1975–76
Philippines Council for Agriculture Resources and Research	198,911		1976–80
Universities	13,634	0.01	
East-West Center (Hawaii)	1,500		1976, 1978
University of Hohenheim (Stuttgart)	4,370		1980
United Nations University	7,764		1980
Others	61,557	0.05	1966, 1969, 1977
Total	127,100,210		

Source: International Rice Research Institute, annual reports, 1962–80.

Table 2.4. Compound rates of growth

	Production		Area		Yield (% per annum)	
Period **Crop**	**(a)** 1949–50 to 1964–65	**(b)** 1967–68 to 1977–78	**(a)** 1949–50 to 1964–65	**(b)** 1967–68 to 1977–78	**(a)** 1949–50 to 1964–65	**(b)** 1967–68 to 1977–78
Food grains	2.98	2.4	1.34	0.38	1.61	1.53
Nonfood	3.65	2.7	2.52	1.01	1.06	1.15
All crops	3.2	2.5	1.6	0.55	1.6	1.4
Rice	3.37	2.21	1.26	0.74	2.09	1.46
Wheat	3.07	5.73	2.7	3.1	1.24	2.53
Pulses	1.62	0.2	1.87	0.75	0.24	0.42

(a) Gleaned from National Council of Agricultural Research, Annual Report 1976, vol. 1, ch. 3, pp. 230–41.
(b) Directorate of Economic Statistics, *Estimates of Area and Production of Principal Crops in India* (1978–1979).

Table 2.5. Imports of food grains in India on government of India account

Year	Quantity in thousand tons
1949	3,765
1950	2,159
1951	4,801
1952	3,926
1953	2,035
1954	843
1955	711
1956	1,443
1957	3,646
1958	3,224
1959	3,868
1960	5,137
1961	3,495
1962	3,640
1963	4,556
1964	6,266
1965	7,462
1966	10,058
1967	8,672
1968	5,694
1969	3,872
1970	3,631
1971	2,054
1972	445
1973	3,614
1974	4,874
1975	7,407
1976	6,483
1977	547

Source: Directorate of Economics and Statistics, New Delhi.

Table 4.1. Cereal consumption (kilogram per capita per year)

Country	2006–7	2007–8	2008–9	2009–10	2010–11
China	382.2	389.1	397.04	411.5	420.9
U.S.	277.6	310.4	325.2	330.4	340.6
India	193.1	197.3	203	207	212

Source: FAO, *Food Outlook* (2010).

Table 4.2. Case study comparing nutritive output per acre for biodiverse vs. monoculture farms in Uttaranchal, India

	Biodiverse	Monoculture
Protein	33 x 8.3 kg	90 kg
Carbohydrate	680 kg	920 kg
Fat	107.8 kg	12 kg
Carotene	2,540 mg	24 mg
Folic acid	554 mg	0
Vitamin C	400 mg	0
Ca	3,420 mg	120 mg
Fe	100.8 g	38.4 g
P	6,013 g	2,280 g
Kg	2,389 g	1,884 g
Na	79 g	0
K	4,272 g	0

Table 5.1. Nutritional content of different food crops

	Protein (g)	Minerals (100 g)	Ca (mg)	Fe (100 g)
Bajra	11.6	2.3	42	5
Ragi	7.3	2.7	344	6.4
Jowar	10.4	1.6	25	5.8
Wheat (milled)	11.8	0.6	23	2.5
Rice (milled)	6.8	0.6	10	3.1

Table 5.2. Some comparatively fast-growing indigenous species

Species	Age (yrs)	MAI (m³/ha)
Duabanga sonneratioides	47	19
Alnus nepalensis	22	16
Terminalia myriocarpa	8	15
Evodia meliafolia	11	10
Michelia champaca	8	18
Lophopetalum fibriatum	17	15
Casuarina equisetifolia	5	15
Shorea robusta	30	11
Toona ciliata	5	19
Trewia nudiflora	13	13
Artocarpus chaplasha	10	16
Dalbergia sissoo	11	34
Gmelina arborea	3	22
Tectona grandis	10	12
Michelia oblonga	14	18
Bischofia javanica	7	13
Broussonatia papyrifera	10	25
Bucklandia populnea	15	9
Terminalia tomentosa	4	10
Kydia calycina	10	11

MAI = mean annual increment

Table 5.3. Yield table for eucalyptus hybrid

Site quality	Age (yrs)	MAI (m³/ha) (OB)	Current AI (m³/ha) (OB)
Good	3	8.1	—
	4	11.3	10.6
	5	13.5	22.3
	6	14.4	18.7
	7	13.9	11.3
	8	13.5	10.6
	9	12.9	8
	10	12.3	6.7
	11	11.6	5.2
	12	11	3.5
	13	10.4	3.6
	14	9.9	3.7
	15	9.4	1.9
Poor	3	0.1	—
	4	0.4	1.4
	5	0.7	1.7
	6	0.8	1.7
	7	0.9	1.2
	8	1	1.4
	9	1	1
	10	1	1.3
	11	1	1.1
	12	1.2	0.7
	13	1	0.8
	14	0.9	0.8
	15	0.9	0.4

AI = annual increment; OB = over bark; MAI = mean annual increment

Table 5.4. Crown biomass productivity of some well-known fodder trees

Species	Crown biomass (tons/ha/yr)
Acacia nilotica	13–27
Grewia optiva	33
Bauhinia	47
Ficus	17.5
Leucena leucocephala	7.5
Morus alba	24
Prosopis sineraria	30

Table 5.5. Grain and straw production of rice varieties

Variety	Grain (lb per acre)	Straw (lb per acre)
Chintamani sanna	1,663	3,333
Budume	1,820	2,430
Halubbalu	1,700	2,740
Gidda byra	1,595	2,850
Chandragutti	2,424	3,580
Putta bhatta	1,695	3,120
Kavada bhatta	2,150	2,940
Garike sanna	2,065	2,300
Alur sanna	1,220	3,580
Bangarkaddi	1,420	1,760
Banku (rainy season 1925–26)	1,540	1,700
G.E.B. (rainy season 1925–26)	1,900	1,540

Table 5.6. Comparison of local and dominant knowledge systems

Local system	Dominant system
Forestry and agriculture integrated.	Forestry separate from agriculture.
Integrated systems have multidimensional outputs. Forests produce wood, food, fodder, water, etc. Agriculture produces diversity of food crops.	Each separate system made one-dimensional. Forests produce only commercial wood. Agriculture produces only commercial crops with industrial inputs.
Productivity in local system is a multidimensional measure that has a conservation aspect.	Productivity is a one-dimensional measure that is unrelated to conservation.
Increasing productivity involves increasing the multidimensional outputs and strengthening the integration.	Increasing productivity involves increasing one-dimensional output by breaking up integrations and displacing diverse outputs.
Productivity based on conservation of diversity.	Productivity based on creation of monocultures and destruction of diversity.
Sustainable system.	Nonsustainable system.

Table 6.1. Essential micronutrients

Vitamin A	Pantothenic acid	Iodine	Manganese	Thiamin
Vitamin D	Vitamin B12	Zinc	Iron	Riboflavin
Vitamin K	Ascorbic acid	Copper	Chromium	Nicotinic acid
Vitamin E	Essential fatty acids (no. 6 & no. 3)	Selenium	Cobalt	Pyridoxine
Folic acid	Biotin			

Table 6.2. Micronutrient-rich foods

Vegetables	Rape leaves, cauliflower greens, amaranth, curry leaves, garden cress, drumstick (leaves), fenugreek seeds, beet greens, purslane, mint, carrots, lotus stems, tapioca chips, colocasia, radishes, sweet potatoes, yams, ivy gourds, lettuce, agathi, radish leaves
Condiments & spices	Poppy, cumin, coriander, oregano, green chilies (fresh/dry), turmeric, ginger, fenugreek, pepper, garlic, mango powder
Nuts & oilseeds	Coconuts (deoiled/dry/milk), groundnuts, cashew nuts, pistachio nuts, gingelly seeds, garden cress seeds, safflower seeds, mustard seeds, niger seeds
Fruits	Indian gooseberries, watermelons, custard apples, wood apples, tomatoes, guavas, mangos, pineapples, oranges, papayas, grapes, baels, pomegranates, gooseberries, apricots

Table 6.3. Comparative study of macronutrients produced per acre of farmland: Mixed cropping vs. monocropping

	Protein (kg)	Carbohydrate (kg)	Fat (kg)	Total energy (kcal)
Mixed cropping				
Mandua = 3 qt	21.9	216	3.9	984,000
Jhangora = 2 qt	12.4	131	4.4	614,000
Gahat = 4 qt	88	228.8	2	1,284,000
Bhatt = 5 qt	216	104.5	97.5	2,160,000
Total = 14 qt	338.3	680.3	107.8	5,042,000
Monocropping				
Paddy = 12 qt	90	920.4	12	4,152,000
Total = 12 qt	90	920.4	12	4,152,000

Sources: Navdanya; ICMR, *Nutritive Value of Indian Foods.*

Table 6.4. Comparative study of vitamins produced per acre of farmland: Mixed cropping vs. monocropping

	Carotene (mg)	Thiamine (mg)	Riboflavin (mg)	Niacin (mg)	Vitamin B6 (mg)	Folic acid (mg)	Vitamin C (mg)	Choline (mg)
Mixed cropping								
Mandua = 3 qt	126	1,260	570	3,300	0	54.9	0	0
Jhangora = 2 qt	0	660	200	8,400	0	0	0	0
Gahat = 4 qt	284	1,680	800	6,000	0	0	400	0
Bhatt = 5 qt	2,130	3,650	1,950	16,000	—	500	—	—
Total = 14 qt	2,540	7,250	3,520	33,700	0	554.9	400	0
Monocropping								
Paddy = 12 qt	24	2,520	1,920	46,800	0	0	0	924,000
Total = 12 qt	24	2,520	1,920	46,800	0	0	0	924,000

Sources: Navdanya; ICMR, *Nutritive Value of Indian Foods.*

Table 6.5. Comparative study of major minerals produced per acre of farmland: Mixed cropping vs. monocropping

	Ca (g)	Fe (g)	P (g)	Mg (g)	Na (g)	K (g)	Cl (g)
Mixed cropping							
Mandua = 3 qt	1,032	11.7	849	411	33	1,224	132
Jhangora = 2 qt	40	10	560	164	0	0	0
Gahat = 4 qt	1,148	27.1	1,244	624	46	3,048	32
Bhatt = 5 qt	1,200	52	3,450	1,190	—	—	—
Total = 14 qt	3,420	100.8	6,103	2,389	79	4,272	164
Monocropping							
Paddy = 12 qt	120	38.4	2,280	1,884	0	0	0
Total = 12 qt	120	38.4	2,280	1,884	0	0	0

Sources: Navdanya; ICMR, *Nutritive Value of Indian Foods.*

Table 6.6. Comparative study of trace minerals produced per acre of farmland: Mixed cropping vs. monocropping

	Cu (mg)	Mn (mg)	Mo (mg)	Zn (mg)	Cr (mg)	S (mg)
Mixed cropping						
Mandua = 3 qt	1,410	16,470	306	6,900	84	480,000
Jhangora = 2 qt	1,200	1,920	0	6,000	180	0
Gahat = 4 qt	7,240	6,280	2,996	11,200	96	724,000
Bhatt = 5 qt	5,600	10,550	—	17,000	140	—
Total = 14 qt	15,450	35,220	3,302	41,100	500	1,204,000
Monocropping						
Paddy = 12 qt	2,880	13,200	936	16,800	108	0
Total = 12 qt	2,880	13,200	936	16,800	108	0

Sources: Navdanya; ICMR, *Nutritive Value of Indian Foods.*

Table 6.7. Comparative study of major macronutrients produced per acre of farmland: Mixed cropping vs. monocropping

	Protein (kg)	Carbohydrate (kg)	Fat (kg)	Total energy (kcal)
Mixed cropping				
Mandua = 6 qt	43.8	432	7.8	1,968,000
Foxtail millet = 3 qt	36.9	182.7	12.9	993,000
French beans = 3 qt	5.1	13.5	0.3	78,000
Amaranth = 2 qt	28	130	14	742,000
Total = 14 qt	113.8	758.2	35	3,781,000
Monocropping				
Paddy = 12 qt	90	920.4	12	4,152,000
Total = 12 qt	90	920.4	12	4,152,000

Sources: Navdanya; ICMR, *Nutritive Value of Indian Foods.*

Table 6.8. Comparative study of vitamins produced per acre of farmland: Mixed cropping vs. monocropping

	Carotene (mg)	Thiamine (mg)	Riboflavin (mg)	Niacin (mg)	Vitamin B6 (mg)	Folic acid (mg)	Vitamin C (mg)	Choline (mg)
Mixed cropping								
Mandua = 6 qt	252	2,520	1,140	6,600	0	109.8	0	0
Foxtail millet = 3 qt	96	1,770	330	9,600	0	45	0	0
French beans = 3 qt	396	240	180	900	0	136.5	72,000	0
Amaranth = 2 qt	—	200	400	1,800	1,200	164	6,000	—
Total = 14 qt	744	4,730	2,050	18,900	1,200	455.3	78,000	0
Monocropping								
Paddy = 12 qt	24	2,520	1,920	46,800	0	0	0	924,000
Total = 12 qt	24	2,520	1,920	46,800	0	0	0	924,000

Sources: Navdanya; ICMR, *Nutritive Value of Indian Foods.*

Table 6.9. Comparative study of major minerals produced per acre of farmland: Mixed cropping vs. monocropping

	Ca (g)	Fe (g)	P (g)	Mg (g)	Na (g)	K (g)	Cl (g)
Mixed cropping							
Mandua = 6 qt	2,064	23.4	1,698	822	66	2,448	264
Foxtail millet = 3 qt	93	8.4	870	243	13.8	750	111
French beans = 3 qt	150	1.83	84	114	12.9	360	30
Amaranth = 2 qt	318	15.2	1,114	6.8	—	1,016	—
Total = 14 qt	2,625	48.8	3,766	1,185.8	92.7	4,574	405
Monocropping							
Paddy = 12 qt	120	38.4	2,280	1,884	0	0	0
Total = 12 qt	120	38.4	2,280	1,884	0	0	0

Sources: Navdanya; ICMR, *Nutritive Value of Indian Foods.*

Table 6.10. Comparative study of trace minerals produced per acre of farmland: Mixed cropping vs. monocropping

	Cu (g)	Mn (g)	Mo (g)	Zn (g)	Cr (g)	S (g)
Mixed cropping						
Mandua = 6 qt	2,820	32,940	612	13,800	168	960,000
Foxtail millet = 3 qt	4,200	1,800	210	7,200	90	513,000
French beans = 3 qt	180	360	60	1,260	18	11,000
Amaranth = 2 qt	1,600	6,800	—	5,800	—	—
Total = 14 qt	8,800	41,900	882	28,060	276	1,484,000
Monocropping						
Paddy = 12 qt	2,880	13,200	936	16,800	108	0
Total = 12 qt	2,880	13,200	936	16,800	108	0

Sources: Navdanya; ICMR, *Nutritive Value of Indian Foods.*

Table 6.11. Time trends in dietary intake and nutritional status of adults: Consumption of different foods (cu/day) vs. Recommended Daily Allowances (RDA)

Food	1975	1980	1990	1995	1996–97	RDA
Cereals and millets (g)	523	533	490	464	450	460
Pulses	32	33	32	33	27	40
Green leafy vegetables	11	14	11	13	15	40
Other vegetables	51	75	49	40	47	60
Fruits	10	25	23	22	—	—
Fats and oils	9	10	13	13	12	20
Sugar/jaggery	19	18	29	23	21	30
Milk & milk products	80	88	96	95	86	150

Table 6.12. Intake of nutrients (cu/day)

Nutrient	1975	1980	1990	1995	1996–97	RDA
Protein (g)	64	52	62	56	54	60
Energy (kcal)	2,296	2,404	2,283	2,172	2,108	2,425
Iron (mg)	32	30	28	26	25 (14*)	30
Vitamin A (eq. mg)	263	313	294	298	282	600
Vitamin B (mg)	0.98	0.91	0.94	0.8	0.9	14
Vitamin C (mg)	41	52	37	35	40	40

Source: Krishnaswamy et al., National Nutrition Monitoring Bureau at the National Institute of Nutrition, India, 1999 report.
*Recently, there has been revision in the iron content of different foods, due to improvements in the procedures of iron estimation. The "revised" iron values are in general less than the "old values." In the present report, data was analyzed, using both values to facilitate comparison with earlier data base. As per the revised values, the intakes are below the RDI in all the states in the current survey. (http://nnmbindia.org/NNMB-PDF%20FILES/Report_OF_2nd%20Repeat_Survey-96-97.pdf)

Table 6.13. Average production of macronutrients per acre of farmland: Organic mixed croppping vs. conventional monocropping

	Protein (kg)	Carbohydrate (kg)	Fat (kg)	Total energy (kcal)
Average production of nutrients from organic mixed farming	240	833	66	4,914,270
Average production of nutrients from conventional monocropping	116	785	23	3,711,475

Table 6.14. Average production of vitamins per acre of farmland: Organic mixed cropping vs. conventional monocropping

	Carotene (mg)	Thiamine (mg)	Riboflavin (mg)	Niacin (mg)	Vitamin B6 (mg)	Folic acid (mg)	Vitamin C (mg)	Choline (mg)
Average production of nutrients from organic mixed farming	2,919	6,550	3,179	31,443	821	878	24,145	680,675
Average production of nutrients from conventional monocropping	745	3,911	1,685	28,381	475	328	36,833	537,527

Table 6.15. Average production of major minerals per acre of farmland: Organic mixed cropping vs. conventional monocropping

	Ca (g)	Fe (g)	P (g)	Mg (g)	Na (g)	K (g)	Cl (g)
Average production of nutrients from organic mixed farming	2,166	82	5,158	1,866	197	6,076	323
Average production of nutrients from conventional monocropping	731	43	3,117	1,496	158	3,465	320

Table 6.16. Cutoff points for diagnosis of anemia

	Hemoglobin (g/dl) in venous blood
Adult males	13
Adult females, nonpregnant	12
Adult females, pregnant	11
Children, 6 months–6 years	11
Children, 6–14 years	12

Table 6.17. Requirement of iron for different age groups

Age group	Iron (mg) that should be absorbed daily
Infants (5–12 months)	0.7
Children (1–12 years)	1.0
Adolescents (13–16 years)	
Males	1.8
Females	2.4
Adult males	0.9
Adult females	
During menstruation	2.8
During first half of pregnancy	0.8
During second half of pregnancy	3.5
While lactating	2.4
After menopause	0.7

Table 6.18. Average production of trace minerals per acre of farmland: Organic mixed cropping vs. conventional monocropping

	Cu (mg)	Mn (mg)	Mo (mg)	Zn (mg)	Cr (mg)	S (mg)
Average production of nutrients from organic mixed farming	12,591	25,124	3,694	43,977	345	1,640,791
Average production of nutrients from conventional monocropping	6,101	15,629	1,077	26,769	157	1,303,224

Table 9.2. Acreage under different varieties of cotton in Warangal, 1996–97

Variety	Hectares	Cost of 450-gram
RCH-2	60,080	250
H-4	2,500	260
NH-44	4,100	250
JKHY-1	3,800	250
Mahyco	8,100	250
Nath	8,200	250
Vanapamula	4,800	250
Others	9,066	250

Source: Office of the Joint Director of Agriculture, Warangal, 1997.

Table 9.3. Cotton arrival and prices in the Warangal agriculture market

Year	Arrival (quintals)	Price per quintal* (Rs.)
1985–86	177,929	437
1986–87	162,332	585
1987–88	608,592	793
1988–89	510,296	786
1989–90	564,290	761
1990–91	432,364	785
1991–92	373,430	1,233
1992–93	572,643	1,040
1993–94	772,999	1,257
1994–95	676,993	1,809
1995–96	1,135,972	1,742
1996–97	1,338,330	1,618
1997–98	833,000	1,800

* Annual average rate per quintal.
Source: Cotton Cooperative Office, Warangal.

Table 9.4. Andhra Pradesh state requirement of seeds

Year	Seed requirement (quintals)
1994–95	955,892
1995–96	985,822
1996–97	1,016,720
1997–98	1,133,205
1998–99	1,378,489
1999–2000	1,756,300

Table 9.5. Public sector seed producers in the state of Karnataka

Company	Seeds
Karnataka State Seed Corporation, Ltd.	Paddy, ragi, maize, bajra, black gram, cowpea, red gram, sunflower, soybean, groundnut, French bean, cotton
National Seeds Corporation, Ltd.	Maize, bajra, paddy, ragi, cowpea, tur, groundnut, soybean, sunflower, jute
University of Agricultural Sciences	Maize, cotton
Karnataka State Department of Agriculture	Bajra, green gram, soybean, tur
State Farms Corporation India, Ltd.	Paddy, maize, jute
Karnataka Oilseed Growers Federation, Ltd.	Paddy, maize, groundnut, soybean, cotton

Table 9.6. Private sector seed producers, per crop

Crop	Companies
Maize	Mahesh Hybrid Seeds; Varada Seeds; Bhadra Hybrid Seeds; Somnath Seeds; Karnataka Hitech Ent; Basaveswara Agro Seeds; Karshek Seeds; Patil Agro; Mahyco; Sumanth Seeds
Cotton	Mahesh Hybrid Seeds; MSSC; Raja Rajeswari Seeds; Ganga Kaveri Seeds; Siddheswara Seeds; Vani Seeds; Sree Hybrid Seeds; Mahantesh Seeds; Rallis Hybrid Seeds; Bhadra Hybrid Seeds; Somnath Seeds; Karnataka Hitech Ent; Nandi Seeds; T S R Amareswara; Amarewara Agri Tech; Sagar Seeds; Laxmi Mills; Vinayaka Agro Seeds; Zauri Seeds; SPIC Bio Tech; Karnataka Seeds; HLL; Mahyco; Vasu & Co; Karnataka Agro Genetics; Mohan Traders; Niranjan Seeds; Mahagujarath Seeds; Adavi Amareswara Seeds; Sumanth Seeds; MHSC; Novarties, Ltd; Laxmi Hybrid Seeds; Rait Hybrid Seeds; Viba Agro Tech; Sri Amarewara Seeds; Manjushree Plantations; Nuziveedu Seeds; Shiva Seeds; T N Amareswara Seeds; Deepthi Seeds; HYCO; Venkateswara Seeds; Advanta; NFCL; Banashankari Seeds; Ashwini Seeds; Kwality Seeds; Shathavahana; Sumantha Hybrid Seeds; Bhubaneswari Seeds; Prabhat Agri Bio-Tech; Amarewari Hybrid Seeds; Kaveri Seeds; Pro Agro Seeds; Pruthivi Agro Tech.
Paddy	Mahesh Hybrid Seeds; Raja Rajeswari Seeds; Ganga Kaveri Seeds; Varada Seeds; Bhadra Hybrid Seeds; Mahyco; Agro Seeds
Bajra	Karnataka Hitech Ent; Sagar Seeds; Karnataka Agro Seeds; CJ Parekh; Mahyco
Sunflower	Sagar Seeds
Tur	Surya Seeds; Agro Seeds

Table 9.7. Distribution of operational holdings by size and area, 1990–91

Holding size	Number	% of holdings	Area operated (ha)	% of total area	Average size of holdings (ha)
Marginal (less than 1 ha)	295,668	26.47	164,224	4.07	0.56
Small (1 to 2 ha)	203,842	18.25	328,215	8.14	1.61
Subtotal (less than 2 ha)	499,510	44.72	492,439	12.21	0.99
Semi-medium (2 to 4 ha)	288,788	25.86	841,541	20.87	2.91
Subtotal (less than 4 ha)	788,298	70.58	1,333,980	33.00	1.69
Medium (4 to 10 ha)	261,481	23.41	1,621,811	40.22	6.2
Large (10 ha and above)	67,171	6.01	1,076, 892	26.7	16.03
Total	1,116,951	100	4,032,683	100	3.61

Source: Statistician Department of Agriculture, Punjab, *Agricultural Statistics of Punjab on the Eve of New Millennium: 2000.*

Table 9.11. Consumption of chemical fertilizers (000s nutrient tons) in Punjab

Fertilizer	1970–71	1980–81	1990–91	1995–96	1996–97	1997–98	1998–99
Nitrogenous	175	526	877	1,020	962	1,005	1,081
Phosphatic	31	207	328	227	229	287	275
Potassic	7	29	15	16	17	22	19
Total (N-P-K)	213	762	1,220	1,263	1,208	1,314	1,375
Consumption per hectare (in kg)	38	14	16	16	15	16	—

Source: Statistician Department of Agriculture, Punjab, *Agricultural Statistics of Punjab on the Eve of New Millennium, 2000.*

Table 9.12. Area under cotton in Punjab

American cotton varieties	Area (000s ha)		
Variety	1998–99	1999–2000	2000–2001
F-846	237.30	151.84	57.9
F-1378	53.10	75.5	131.92
LH-1556	37.50	26.5	79.8
Hybrids	10.20	31.13	76.82
Skinderpuri	32.30	20.4	29.8
F-414	34	20	14.85 (Pk 54)
Others	47.90	43.93	48.82
Total	*452.30*	*369.30*	*439.91*
LD-327		560.4	82.19
RG-8		433.1	24.68
Others		12.2	3.22
State Total	*562*	*475*	*550*

Source: Department of Agriculture, Punjab.

Table 9.13. Details of the use of chemical pesticides in cotton cultivation in Bhatinda, Punjab.

Ist spray of systemic products   *60 days after planting (one among the six given below)*

Systemic products	Dosage/Acre	Price/Acre
Confidor	40 ml	Rs. 125
Monocrotophos	400 ml	Rs. 100
Metasistocs	400 ml	Rs. 125
Roger	400 ml	Rs. 80
Dymecon	150 ml	Rs. 60
Endosalpha	1 l	Rs. 200

Dosage of the first spray recommended by Department of Agriculture, Bhatinda
2nd spray of systemic products   *70–75 days after planting (in the same dosages)*
3rd spray of synthetic pyrethoid   *80–90 days after planting (one among the five given below)*

Synthetic pyrethroid	Dosage/Acre	Price/Acre
Fenvalerate	150 ml	Rs. 40
Cypermathrin	200 ml	Rs. 50
Alphamethrin	150 ml	Rs. 70
Karate	150 ml	Rs. 70
Decameterin	200 ml	Rs. 100

4th spray is often a cocktail of one of the synthetic pyrethoids mixed with one of the three given

Chemical name	Dosage/Acre	Price/Acre
Cloropariphos	1 l	Rs. 150
Etheon	800 ml	Rs. 150
Quienalphos	800 ml	Rs. 160

On average, 9–10 sprays are made on cotton crops in the entire cotton belt in Punjab, including in Bhatinda. Sometimes there are as many as 15. After the 4th spray, farmers simply make a cocktail of chemicals from the 1st, 3rd, and 4th sprays, choosing the chemicals randomly. Sometimes two chemicals are used, but very often, as witnessed during the heavy pest infestations of the last few years, 3–4 chemicals are mixed.

Table 9.14. Cropping patterns in Punjab (%)

Crop	1950–51	1960–61	1970–71	1980–81	1990–91	1995–96	1996–97	1997–98
Paddy	2.9	4.8	6.9	17.5	26.9	28.2	27.7	29.1
Maize	6.3	6.9	9.8	5.6	2.5	2.2	2.1	2.1
Bajra	5.2	2.6	3.7	1	0.2	0.1	0.1	0.1
Wheat	27.3	29.6	40.5	41.6	43.6	41.6	41.43	42.1
Barley	2.4	1.4	4	0.9	0.5	0.5	0.4	0.5
Total pulses	23.8	19.1	7.3	5	1.9	1.3	1.3	1.1
Total oilseeds	3.3	3.9	5.2	3.7	1.5	3	3.2	2.5
Sugarcane	2.2	2.8	2.3	1	1.3	1.8	2.2	1.6
Cotton	5.4	9.4	7	9.6	9.3	9.6	9.5	9.2
Total vegetables	1.2	1.2	0.9	1.1	0.7	1	1	1.1
Total fruits	0.8	0.6	0.6	0.4	0.8	1.1	1.1	1.1
Other crops	19.2	17.7	14.8	12.6	10.8	9.6	10	9.5
Total cropped area	100	100	100	100	100	100	100	100

Source: Statistician Department of Agriculture, Punjab, *Agricultural Statistics of Punjab on the Eve of New Millennium: 2000.*

Table 9.15. Agricultural implements and machinery in use in Punjab (000s)

	1995	1996	1997	1998	1999
Tractors/trailers	320	330	350	365	375
Tillers/cultivators	220	228	235	245	250
Disc harrows (tractor drawn)	240	248	255	265	265
Seed-cum-fertilizer drills	130	135	140	145	155
Spray pumps	485	510	525	540	545
Tractor-drawn combines	4.4	4.6	4.7	4.8	4.9
Self-propelled combines	2.2	2.3	2.4	2.5	2.7
Threshers	305	305	315	325	340
Cane crushers	35	35	35	35	30
Tube wells	860	875	900	925	935

Source: Statistician Department of Agriculture, Punjab, *Agricultural Statistics of Punjab on the Eve of New Millennium: 2000.*

Table 9.16. Advancement of credit to farmers (Rs. in crores)

	1990–91	1995–96	1996–97	1997–98	1998–99
Kharif season	159.01	440.86	562.57	693.63	804.31
Cash	79.25	262.75	371.92	468.92	548.28
Kind	79.76	178.11	190.65	224.71	256.03
Rabi season	204.12	505.65	535.60	679.28	898.69
Cash	102.83	232.98	274.46	364.46	564.72
Kind	101.29	272.67	261.14	314.82	333.97
Total	363.13	946.51	1,098.17	1,372.91	1,703.00
Cash	182.08	495.73	646.38	833.38	1,113.00
Kind	181.05	450.78	451.79	539.53	590.00

Source: Statistician Department of Agriculture, Punjab, *Agricultural Statistics of Punjab on the Eve of New Millennium: 2000.*

Table 9.17. Farmer suicides in all of India, 1995–2012

Year	Number of suicides
1995	10,720
1996	13,729
1997	13,622
1998	16,015
1999	16,082
2000	16,603
2001	16,415
2002	17,971
2003	17,164
2004	18,241
2005	17,131
2006	17,060
2007	16,632
2008	16,196
2009	17,368
2010	15,964
2011	14,027
2012	13,754
Total	**284,694**

Table 9.18. Farmer suicides in Maharashtra, 1995–2012

Year	Number of suicides
1995	1083
1996	1981
1997	1917
1998	2409
1999	2423
2000	3022
2001	3536
2002	3695
2003	3836
2004	4147
2005	3925
2006	4453
2007	4238
2008	3802
2009	2872
2010	3141
2011	3337
2012	3786

Source: National Crime Records Bureau

Table 9.19. Farmer suicides in Vidarbha, 2001–2013

Year	Number of suicides
2001*	52
2002	104
2003	148
2004	447
2005	445
2006	1448
2007	1247 (350) = 1592
2008	1148 (340) = 1488
2009	1005 (273) = 1278
2010	1177 (279) = 1456
2011	999 (358) = 1357
2012	950 (425) = 1375
2013	752 (257) = 1009

Note: Numbers in parentheses represent farmer-related suicides.
Source: Deshpande, 2014.
* For 2001–2006 (Maitra, 2012; Baweja, 2012)

Table 9.20. Farmers who committed suicide during November–December 1998 in Warangal district

Farmer	Age	Village	Mandal	Date of suicide
Ketapalli Sambi Reddy	40	Ogalpur	Atmakur	10-22-98
Bhukya Sarma	35	Harischandra Nayak Tandra	Hasanparti	11-8-98
Kari Kumari Lingayya	49	Gidde Muttaram	Chiyala	11-11-98
Malotu Danja	40	Mangalvaripeta	Khanapuram	11-12-98
Nagelli Tirupati Reddy	26	Challagange	Chityala	11-14-98
Indla Ayilayya	36	Neredupalli	Bhupalapalli	11-18-98
Pacchi Kalaya Someswara	48	Aakinepalli	Mangapeta	11-19-98
Kattula Yakayya	32	Samudrala	Stn Ghanpur	11-19-98
Akutota Venkatayya	65	Govindapuram	Sayampeta	11-21-98
Bolla Hari Krishna	22	Nadikuda	Parakala	11-24-98
Edelli Lakshmi	45	Rauvlapalli	Regonda	11-18-98
Cheviti Veeranna	28	Tehsildar Banjar	Dornakal	12-3-98
Pentla Odelu	42	Nagurlapelli	Regonda	12-16-98
Ragula Devender Reddy	25	Jubilee Nagar	Regonda	12-16-98
Tallapalli Lakshmayya	38	Solipuram	Narmetta	12-18-98

Source: Prajasakhti newspaper.

Table 9.21. Farmers who committed suicide during 1999–2000 in Andhra Pradesh

Farmer	Age	Village	Mandal	District
Bhubanagiri John Reddy	40	Gannevaram	Yedanpudi	Prakasham
Ravipati Koteswar Rao	37	Poluru	Yedanpudi	Prakasham
Gogati Bali Reddy	—	Kuntalapalli	Nallamada	Ananthapur
Kalmula Ramayya	60	Macharam	Amrabad	Mehboobnagar
Pallepu Ankamma	45	Paladugu	Medikonduru	Guntur
Kethavathrathan	30	Inumulanarva	Kotthur	Mehboobnagar
Yadayya	28	Rajapuram	Balanagar	Mehboobnagar
Boya Pengayya	—	Gangapuram	Zedcherla	Mehboobnagar

Source: Rathuy vani, various issues.

Table 12.1. Fertility increase due to organic farming

	Percent increase over industrial farming						
Crop	Organic matter	Microbial activity	Microbial biomass	Water-holding capacity	N	P	K
Pearl millet	28–55	4–25	2–10	2–3	0–2	0–1	8–15
Cluster bean	32–44	22–54	12–25	4–9	12–34	2–4	25–47
Moth bean	31–47	11–23	8–15	4–7	7–21	1–2	4–9
Mung bean	27–41	28–59	11–33	4–8	11–27	2–6	5–11

Note: Results are consolidated from farms where organic farming has been practiced for four or more years.

Table 12.2. Biodiverse vs. monoculture production

Crop	Avg. production (kg/ha)	Avg. price (per kg)	Total income (Rs.)
Baranaja			
Bajra	440	8	3,520
Maize	1,280	8	10,240
Safed chemi	600	25	15,000
Ogal	360	20	7,200
Mandua	600	10	6,000
Jhangora	440	15	6,600
Urad	600	20	12,000
Navrangi	680	20	13,600
Koni no. 1	280	10	2,800
Lobia	600	20	12,000
Til	400	30	12,000
Koni no. 2	340	10	3,400
Total	6,620		104,360
Monoculture			
Maize	5,400	8	43,200
Navdanya			
Til	400	30	12,000
Safed chemi	720	25	18,000
Mandua	1,120	10	11,200
Dholiyia dal	640	20	12,800
Safed bhatt	760	15	11,400
Lobia	800	20	16,000
Jhangora	520	15	7,800
Maize	560	8	4,480
Wheat	480	25	12,000
Total	6,000		105,680
Monoculture			
Mandua	3,600	10	36,000
Saptarshi			
Urad	600	20	12,000
Moong	520	25	13,000
Mandua	560	10	5,600
Safed bhatt	680	15	10,200
Dholiyia dal	560	20	11,200
Maize	680	8	5,440
Lobia dal	600	20	12,000
Total	4,200		69,440
Monoculture			
Urad	2,400	20	48,000

Source: Sir Albert Howard, *An Agricultural Testament* (Goa: Other India Press / RFSTE, 2000), 13.

Table 13.1. World's top ten seed companies

Company	2007 seed sales (U.S. $ millions)	% of global proprietary seed market
Monsanto (U.S.)	4,694	23
Dupont (U.S.)	3,300	15
Sygenta (Switzerland)	2,018	9
Groupe Linagrain (France)	1,226	6
Land O'Lakes (U.S.)	917	4
KWS AG (Germany)	702	3
Bayer Crop (Germany)	524	2
Sahata (Japan)	396	< 2
DLF Trifolum (Denmark)	391	< 2
Takii (Japan)	347	< 2
Total	14,785	67

Source: ETC, *Who Owns Nature*, http://etcgroup.org/upload/publication/707/01/etc_won_report_final color.pdf.

Selected Bibliography

Achebe, Chinua. *No Longer at Ease.* London: Heinemann, 1960.

Anderson, Robert, and Baker Morrison. *Science, Politics and the Agricultural Revolution in Asia.* Boulder, CO: Westview, 1982.

Bailey, Britt, and Marc Lappe. *Against the Grain: Biotechnology and the Corporate Takeover of Your Food.* Monroe, ME: Common Courage, 2002.

Belcher, Brian, and Geoffrey Hawtin. *A Patent on Life: Ownership of Plant and Animal Research.* Canada: IDRC, 1991.

Caufield, C. *In the Rainforest.* London: Picador, 1986.

Chesler, Phyllis. *Sacred Bond: Motherhood under Siege.* London: Virago, 1988.

Clarke, Robin. *Water: The International Crisis.* Cambridge, MA: MIT Press, 1993.

Coats, David. *Old McDonald's Factory Farm.* New York: Continuum, 1989.

Collier, Paul. *The Plundered Planet.* Oxford: Oxford University Press, 2010.

Custers, Peter. *Women in the Tebhaga Uprising.* Calcutta: Naya prokash, 1987.

Darwin, Charles. *The Formation of Vegetable Mould through the Action of Worms, with Observations on Their Habits.* London: Faber and Faber, 1927.

Dawkins, Richard. *The Selfish Gene.* Oxford: Oxford University Press, 1976.

Deb, Debal. *Industrial versus Ecological Agriculture.* New Delhi: Navdanya, 2004.

De Villiers, Marq. *Water: The Fate of Our Most Precious Resource.* New York: Houghton Mifflin, 2000.

DeVries, J., and G. Toenniessen. *Securing the Harvest: Biotechnology, Breeding and Seed Systems for African Crops.* Oxfordshire, UK: CABI, 2001.

Doyle, Jack. *Altered Harvest.* New York: Viking, 1985.

Gandhi, M. K. *Food Shortage and Agriculture.* Ahmedabad: Najivan, 1949.

Ghosh, Kali Charan. *Famines in Bengal, 1770–1943.* Calcutta: Indian Associated Publishing, 1944.

Gleaser, B., ed. *The Green Revolution Revisited.* Boston: Allen and Unwin, 1956.

Greenpeace. *Eating Up the Amazon.* Netherlands: Greenpeace, April 2006.

Harding, S. *The Science Question in Feminism.* Ithaca: Cornell University Press, 1986.

Ho, M. W., and P. T. Saunders, eds. *Beyond Neo-Darwinism: Introduction to the New Evolutionary Paradigm.* London: Academic, 1984.

Hong, E. *Natives of Sarawak*. Malaysia: Institut Masyarakat, 1987.

Howard, Albert. *An Agricultural Testament*. London: Oxford University Press, 1940.

Howard, Louise E. *Sir Albert Howard in India*. London: Faber and Faber, 1953.

Hyams, E. *Soil and Civilisation*. London: Thames and Hudson, 1952.

Jackson, Wes et al., eds. *Meeting the Expectations of the Land*. San Francisco: North Point, 1984.

Johnson, A. S. *Cropping Patterns in India*. New Delhi: ICAR, 1978.

Jones, Andy. *Eating Oil: Food in a Changing Climate*. London: Sustain/ELM Farm Research Center, 2001.

Kay, Lily E. *The Molecular Vision of Life: Caltech, the Rockefeller Foundation and the Rise of the New Biology*. Oxford: Oxford University Press, 1993.

Kimbrell, Andrew. *The Human Body Shop*. New York: Harper-Collins, 1993.

Kloppenburg, Jack. *First the Seed*. Cambridge: Cambridge University Press, 1988.

Kovda, V. A. *Land Aridization and Drought Control*. Boulder, CO: Westview, 1980.

Kuhn, T. *The Structure of Scientific Revolutions*. Chicago: University of Chicago Press, 1972.

Lewontin, Richard. *The Doctrine of DNA*. New York: Penguin Books, 1993.

Maturana, Humberto R., and Francisco J. Varela. *The Tree of Knowledge: The Biological Roots of Human Understanding*. Boston: Shambala, 1992.

Merchant, Carolyn. *The Death of Nature: Women, Ecology and the Scientific Revolution*. New York: Harper and Row, 1980.

Mies, Maria, ed. *Women: The Last Colony*. London: Zed Books, 1989.

Morgan, Dan. *Merchants of Grain*. New York: Viking, 1979.

Nandy, A., ed. *Science, Hegemony and Violence*. Delhi: Oxford University Press, 1988.

Palmer, Tim. *Endangered Rivers and the Conservation Movement*. Berkeley: University of California Press, 1986.

Panday, K. K. *Fodder Trees and Tree Fodder in Nepal*. Berne: Swiss Development Cooperation, 1982.

Park, K. *Park's Textbook of Preventive and Social Medicine*. 21st ed. India: Banarsidas Bhano, 2011.

Pearse, Andrew. *Seeds of Plenty, Seeds of Want*. Oxford: Oxford University Press, 1980.

Peat, M. M., and I. D. Teare. *Crop-Water Relations*. New York: Wiley, 1983.

Pfeiffer, Dale Allen. *Eating Fossil Fuels: Oil, Food, and the Coming Crisis in Agriculture*. Gabriola Island, BC: New Society, 2006.

Pilger, John. *A Secret Country.* London: Vintage, 1989.

Pollan, Michael. *The Omnivore's Dilemma: A Natural History of Four Meals.* New York: Penguin, 2006.

Postman, Neil. *Technology: The Surrender of Culture to Technology.* New York: Knopf, 1992.

Randhawa, M. S. *A History of Agriculture in India.* New Delhi: Indian Council of Agricultural Research, 1989.

Rissler, Jane, and Margaret Mellon. *The Ecological Risks of Engineered Crops.* Cambridge, MA: MIT Press, 1996.

Shah, C. H., ed. *Agricultural Development of India: Policy and Problems.* Delhi: Orient Longman, 1979.

Shiva, Vandana. *Biodiversity Based Organic Farming: A New Paradigm for Food Security and Food Safety.* New Delhi: Navdanya, 2006.

———. *Staying Alive: Women, Ecology, and Development.* London: Zed Books, 1988.

———. *Tomorrow's Biodiversity.* New York: Thames and Hudson, 2000.

———. *The Violence of the Green Revolution: Third World Agriculture, Ecology, and Politics.* London: Zed Books, 1991.

Shiva, Vandana, J. Bandyopadhyay, and H. C. Sharatchandra. *The Social, Ecological and Economic Impact of Social Forestry in Kolar.* Bangalore: IIM, 1981.

Shiva, Vandana, and Gurpreet Karir. *Chemmeenkettu.* New Delhi: Research Foundation for Science, Technology and Ecology, 1997.

Shiva, Vandana et al. *Biosafety.* Penang: Third World Network, 1996.

———. *Ecology and the Politics of Survival: Conflicts over Natural Resources in India.* New Delhi: Sage, 1991.

———. *Monocultures of the Mind: Perspectives on Biodiversity and Biotechnology.* London: Zed Books, 1993.

———. *Seeds of Suicide.* New Delhi: Research Foundation for Science, Technology, and Ecology, 2001.

Singer, Peter, and Deane Wells. *The Reproductive Revolution: New Ways of Making Babies.* Oxford: Oxford University Press, 1984.

Singh, R. V. *Fodder Trees of India.* New Delhi: Oxford University Press, 1982.

Subramaniam, C. *The New Strategy in Agriculture.* New Delhi: Vikas, 1979.

Swaminathan, M. S. *Science and the Conquest of Hunger.* Delhi: Concept, 1983.

Taittreya Upanishad. Gorakhpur: Gita.

Troup, R. S. *Silviculture Systems.* Oxford: Oxford University Press, 1916.

von Weizsäcker, Ernst Ulrich, Amory Lovins, and Hunter Lovins. *Factor Four: Doubling Wealth, Halving Resource Use.* London: Earthscan, 1997.

Waring, Marilyn. *If Women Counted.* New York: Harper and Row, 1988.

Weigle, Marta. *Creation and Procreation.* Philadelphia: University of Pennsylvania Press, 1989.

Wesson, Robert. *Beyond Natural Selection.* Cambridge, MA: MIT Press, 1993.

Wright, Angus. *The Death of Ramon Gonzalez: The Modern Agricultural Dilemma.* Austin: University of Texas Press, 1990.

Copyrights and Permissions

"The Gendered Politics of Food" was originally published in Vandana Shiva, introduction to *Staying Alive: Women, Ecology, and Development* (London: Zed Books, 1988), xiv–xx (first published in India and South Asia by Kali for Women: New Delhi, 1988). Used by permission.

"Science and Politics in the Green Revolution" was originally published in Vandana Shiva, "Science and Politics in the Green Revolution," in *The Violence of the Green Revolution: Third World Agriculture, Ecology, and Politics* (London: Zed Books, 1991), 19–60. Used by permission.

"The Hijacking of the Global Food Supply" was originally published in Vandana Shiva, "The Hijacking of the Global Food Supply," in *Stolen Harvest: The Hijacking of the Global Food Supply* (Cambridge, MA: South End, 2000), 5–20. Used by permission.

"Hunger by Design" was originally published in Vandana Shiva, "Hunger by Design," in *Making Peace with the Earth: Beyond Resource, Land, and Food Wars* (New Delhi: Women Unlimited: An Associate of Kali for Women, 2012; also published North Melbourne, Vic.: Spinifex, 2012). Used by permission.

"Monocultures of the Mind" was originally published in Vandana Shiva, "Monocultures of the Mind," in *Monocultures of the Mind: Perspectives on Biodiversity and Biotechnology* (London: Zed Books, 1993), 9–64. Used by permission.

"Toward a New Agriculture Paradigm: Health per Acre" was originally published as *Health per Acre/Wealth per Acre: A New Paradigm of Productivity*, report (New Delhi: Navdanya).

"Can Life Be Made? Can Life Be Owned? Redefining Biodiversity" was originally published in Vandana Shiva, "Can Life Be Made? Can Life Be Owned? Redefining Biodiversity," in *Biopiracy: The Plunder of Nature and Knowledge* (Boston: South End, 1997), 19–42. Used by permission.

"The Seed and the Earth" was originally published in Vandana Shiva, "The Seed and the Earth," in *Biopiracy: The Plunder of Nature and Knowledge* (Boston: South End, 1997), 43–64. Used by permission.

"Seeds of Suicide" was originally published in Vandana Shiva, Asfar H. Jafri,

333

Index

Culture of the Land: A Series in the New Agrarianism

This series is devoted to the exploration and articulation of a new agrarianism that considers the health of habitats and human communities together. It demonstrates how agrarian insights and responsibilities can be worked out in diverse fields of learning and living: history, science, art, politics, economics, literature, philosophy, religion, urban planning, education, and public policy. Agrarianism is a comprehensive worldview that appreciates the intimate and practical connections that exist between humans and the earth. It stands as our most promising alternative to the unsustainable and destructive ways of current global, industrial, and consumer culture.

Series Editor

Norman Wirzba, Duke University, North Carolina

Advisory Board

Wendell Berry, Port Royal, Kentucky
Ellen Davis, Duke University, North Carolina
Patrick Holden, Soil Association, United Kingdom
Wes Jackson, Land Institute, Kansas
Gene Logsdon, Upper Sandusky, Ohio
Bill McKibben, Middlebury College, Vermont
David Orr, Oberlin College, Ohio
Michael Pollan, University of California at Berkeley, California
Jennifer Sahn, *Orion* Magazine, Massachusetts
Vandana Shiva, Research Foundation for Science, Technology & Ecology, India
Bill Vitek, Clarkson University, New York

CPSIA information can be obtained at www.ICGtesting.com
Printed in the USA
BVOW01s0544200315

392564BV00002B/3/P